Back in Keith County

Back in Keith County

JOHN JANOVY, JR.

University of Nebraska Press
Lincoln and London

Manufactured in the United States of America

First Bison Book printing: 1984
Most recent printing indicated by the first digit below:
1 2 3 4 5 6 7 8 9 10

Library of Congress Cataloging in Publication Data
Janovy, John.
Back in Keith County.
Reprint. Originally published: lst ed. New York, N.Y. : St. Martin's Press, c1981.
1. Janovy, John. 2. Natural history—Nebraska—Keith County. 3. Biologists—United States—Biography. 4. Keith County (Neb.) I. Title.
[QH105.N2J34 1984] 508.782'89 83-17003
ISBN 0-8032-7560-9 (pbk.)

All illustrations in this book are by, and from the collection of, the author.

To those with broad niches
and gifts of survival

Contents

Back in Keith County

1

Erma's Desire

THIS book concerns mainly cowboy country and some things I learned out there—things about how to live and how not to live, things about the environment in which my mind operates. There are a lot of cowboys, homesteaders, and scruffians in this book, along with some of my very best friends (some of whom are pretty scruffy), but they all have one thing in common: they all like to think they live "the good life." Their lives may be "good," but those lives are good often in ways they take for granted. Cowboy country has a way of altering your mind so that all sorts of things begin to be taken for granted. But *Erma's Desire* is not one of these.

She stands off I-80 to the south near a town called Grand Island. Grand Island was so named because it's built partly upon an island in the Platte River. There are times when you could say the Platte is all island and no river, and then wonder about the mental state of those blistered pioneers who arrived at the Platte with tired oxen and decided there was not only a river, but an island. Modern man has laid down a double ribbon of concrete and steel-reinforced bridges—a ribbon that follows the Platte for miles and miles out across Colorado. Going west from the Midwest along I-80, the river and interstate converge at the town of Grand Island, and there sits

Erma's Desire off to the south, hardly visible from the westbound lane unless you know where to look. And I know where to look. She nods to me as I head west, a standard, soft greeting, the slightest of gestures conveying total approval. The Platte passes for an instant beneath my feet with the roar of a bridge crossed. The river is then off to my left, a comforting sight.

Heading back east, however, back to the state capital, into the jaws of paperwork, she speaks to me in a different tone. Her message is sharp, ripping, contemptuous. Going east, the river and highway diverge and I am left alone in an ancient automobile, perhaps the last of its species. A longnosed Kenworth, brand-new, total drama on the highway, never slows, coming around now on the left, now gone into the traffic miles ahead. I can't watch the semi; Erma is on my right. She is only sculpture, I keep telling myself. She is only Cor-ten steel, piled into an array of points and angles. But her contempt for me is brazen and real as I head east. She does not lower herself to actually tell me it's a mistake to leave the West. She doesn't need to; I know it already.

There are three sharp spires to Erma, set at angles on trianguloid bases. The spires make different angles with the horizon, face different directions of the compass, cast reflections of different lengths in the small lake near where Erma sits, apart from the rest stop buildings. The distance from the buildings must have some meaning, almost as if she had placed herself there, away from the pet exercise area, the drinking fountain, the rest rooms, the picnic tables, the sum totality of trophic behavior. Visitors to the plains, surprised to find her there, wander over, walk slowly around and around, each with his own thoughts of what she is and what she means. Visitors to the plains, finished with Erma, leave Grand Island never, perhaps, to return, and start east, asking one another how far it is now to Omaha, Des Moines, Chicago. They have missed it all. One must not stop, walk, look, touch, then leave. One must repeatedly pass, at highway speeds, after having touched her once, going east. She speaks more strongly then. The repeated messages finally make it through a thick skull. The points are, they must be, her desires. They are sharp, as desires must be. The largest one, set at the most insistent angle, points back west, back to Keith County.

Shapes, colors, and movements evoke emotions. If they did not, there would be no painting, no sculpture, no courtship between animals, no territorial defense, no behavioral communications. I have no knowledge of the inner workings of abstract artists' minds, but I do know that the abstract artist routinely extracts from subjects those elements of pure color, pure shape, pure movement, that evoke emotion. Furthermore, in the case of abstract sculpture, the best ones are protean, self-generating, always the same but never the same. Different emotions are evoked at different angles or from the same angle on different days, depending on cloud patterns, wind blowing through spaces, or the color of sunrise. The best ones are thus free of restrictions and the humans who view them are also free of restrictions, free to interpret as they wish, free to feel as they wish, free to think their own thoughts about this pile of steel and free to change those thoughts. There is an independence forced upon us by abstract sculpture, especially by sculpture as large and powerful as *Erma's Desire.* Even when told how to interpret the artist's work, we cannot all agree with that interpretation. No matter how hard we resist, we cannot but think our own thoughts about *Erma's Desire.* Such a gift, artist John Raimondi, such a gift to the people of the interstate!

The history of this country is pretty simple and it all goes like this: some years ago people left the places they were and moved west into uncertainty, riches and poverty, feast and famine, thirst, hunger, hostile natives, dry land, dust, and winter such as few beyond the Blackfoot Indians had ever experienced. If we disregard the fact that these people ended up replacing a native population of humans, who lived an idyllic and harmonious existence with nature, with a population of ranchers, farmers, and center pivot sprinklers, then we can even admire the pioneer spirit. Try as I might, however, I cannot envision those thoughts that must have been in the minds of mothers, young wives, teenage girls, as they prepared to join their men on a walk across Nebraska. There had to be some regrets, some feelings for the places, the friends, and relatives they were leaving. *Erma's Desire* has a spire angled back toward the east, back toward friends and relatives. There must have been some deep religious feelings, some trust and faith in a Higher Power, a Power that would walk with them along the sands of the Platte River. Erma's eastern spire is set at a high angle, pointing not only east, but upward into the heavens. There must have been some overriding determination, some commitment to a totally new and unknown horizon, that in the end took charge of the psyche and powered these people on toward their destinations. West; Erma's strongest spire points west, at a low angle, heavy and sharp, west along the Platte and west into unknown territory.

As I travel east along the interstate this whole message is told again and again to me by *Erma's Desire:* keep your ties with security, with the east, but never let them suppress your determination to move into the unknown, into that world in which your own skills and your own decisions govern your fate. The movements of her spires as I drive east give me this message strongly and sharply. In motion, as I move at highway speed, her eastern spire, her faith and remembrances, remains almost stationary, steadfast. At highway speed her western spire drives powerfully into the unknown, a message that lasts the miles home and then some: Your best place is along the Platte; don't forget it. I won't, Erma, I mumble in the dark out on the interstate. I won't forget it. My best place *is* along the Platte where in spring the wildest call of all the world, the gutteral resonance of sandhill cranes, takes me back to the freedom of primeval times, where in the winter my fish nets freeze solid and bald eagles watch from trees near the interstate, where only my obligations to my own thoughts and ideas guide my eyes and feet.

Not many places in the world where that can happen, are there. Not many places in the world where a person can choose—even if it be for such a short time—to say, to think, and feel only the thoughts of one's own unique mind, with perhaps a companion who's seen *Erma's Desire* and has also his or her unique thoughts about a piece of sculpture that cannot be ignored. I think back over all the years of trying to generate new ideas in this area of original science that is my profession and I can remember well when the new ideas first appeared in conversation. Those times occurred when either I, or some bright student, was freed of every obligation but the obligation to our own thoughts. Those times occurred when we breathed the air of total freedom, including the freedom to express those thoughts knowing full well someone would listen, take us seriously, no matter what set of scientific rules or precedents were violated. At those times the ideas moved about one another, pointed here and there, assumed positions of relative importance that depended on their angle of view, a shifting, changing, intellectual experience, from which a single message finally emerged, at least for the time. The totality of freedom of uninhibited intellectual exchange, no jockeying for positions of power, the flow of unhindered thoughts open to all interpretations, these things remind me of only *Erma's Desire*. The combination of points and directions, emphases and bases, settings and an infinite number of ephemeral contexts, from which finally emerges the luxury of a new idea, a single message—these things say *Erma's Desire* to me. And conversely, *Erma's Desire* has come to signify these things as I head west into that territory where there are no telephones, no forms to complete, no deadlines except the ones imposed by the lives of animals I wish to study.

This then, John Raimondi, is the symbol you have given us, regardless of your original intent. Your pile of Cor-ten steel, stacked out where the river meets the road, is the result of your own unique thinking, and for that reason I value it all the more. Your pile of steel, stacked right out there for all to see from any angle, is the result of your own freedom to bare your ideas. And the public controversy that surrounded Erma when she first appeared in our midst was a manifestation of her power: the power to compel an individual thought, an individual interpretation. Such a gift to the people of the plains, John Raimondi, such a gift for cowboy country! Such an expression of intellectual courage and a contempt for those who choose not to have it!

These are my interpretations. They may have nothing to do with the thoughts that were in your mind when you built her, but I know they are shared by at least one other person on this earth. That other person is one of the most creative, untamed, courageous, independent people who has ever lived on this planet: my oldest daughter. I discovered that out here in cowboy country, and upon making that discovery, decided how to write this book. Everything was revealed one night when my plane landed on an airstrip alongside the Platte River. The storms were on the ocean that night; the plane had been wrenched almost from the sky. But I stepped off the ramp with a boot onto the soil of cowboy country. I had just come from Switzerland.

2

A Cowboy in Switzerland

I have a lot of envious friends, mainly because they don't know any better, but also because of this letter I received quite unexpectedly. I walked into my place of business on a rainy day in September and discovered an airmail letter, flimsy, you know the type, designed especially for *foreign* correspondence. I get very touchy of *foreign* correspondence. It always seems to portend something I really don't want to get involved in, or be responsible for, usually in a language I don't understand, and often in a value set in which I can't participate. Foreign correspondence also gives me the wanderlust and irresponsible itches.

This particular letter was the foreign correspondence of all foreign correspondence. It contained an out-of-the-blue invitation to come to Geneva, at World Health Organization expense, and attend a scientific conference to begin developing solutions to the world's tropical disease problems. In common words, they wanted me on a committee and the committee met in Geneva. They sent me an authorization and a ticket, handed me an expense check when I got there, and treated me like a king, but subsequently made me chairman of another committee. My first committee was a lark. The second was an intellectual disaster that up to this time I have not shared formally with anyone. The first committee I left with a sense of hope, thrilled, with optimism in my heart, confident in my image of a better future for those millions of people throughout the tropics who are being eaten alive by worms. The second I left with a profound sense of concern over the lives of my fellow American countrymen, hardly any of whom ever knew a tropical disease, much less caught one.

I discovered in Geneva that I'd been invited to WHO for a variety of reasons, all centered around the fact that for many years we, meaning some exceedingly bright students and post-docs and me, had studied the functional biology of some one-celled animals. These animals were of a variety of species, most of which were causative organisms of human infections in exotic parts of the world. In general, over the years we'd built a fair reputation based upon our research into the lives of these animals. I also discovered in Geneva that not everyone in the world knew where Nebraska was, or cared. But most significantly, I learned that the way tropical diseases are studied in Nebraska is a far cry indeed from the way tropical diseases are studied in the tropics. Those discoveries caused me untold personal agony over the next few months and they still bother me, seriously, whenever I allow myself to think about them. Those discoveries led me to places of the mind that a member of a highly organized society should probably not be shown. And they led me to conclusions about the ways we live that have absolutely nothing to do with tropical diseases, research, or science, but are only manifest for me to see in those areas.

It is a matter of public record that as a sequela of my first visit to Geneva, WHO gave me a gigantic amount of research money to spend. But it is not a matter of public record how all that came about. Nor are the events that happened at the University of Nebraska, and that led to my receipt of all that money, particularly widely known. There is no reason for them to be widely known, or even considered unusual. They are commonplace daily business and there are people getting paid a bunch of money to make sure this kind of business gets done easily. The events can be told quite simply: I wrote a proposal for some world money, the proposal was sent by the office that sends such documents, it was reviewed by a panel of anonymous experts selected by WHO, and funded. WHO sent a check to my place of business and I began some research on the project. In a few months I sent WHO a progress report. It was all that easy. "We didn't win no Nobel Prize with that progress report," I once said to a couple of cowboys, nor will we win one with the research. But we will eventually get some publications to enhance our scientific reputations. And, I feel the ideas in that proposal were basically good ones. So this is about how science is conducted in the United States of America, even in Nebraska.

There is untold personal freedom that underlies the manner in which this kind of business is conducted throughout this nation. I believe that such untold freedom extends throughout many walks of our lives in ways we never suspect until we are brought up short against situations, circumstances, and environments in which there is no word for "freedom" as *we* know and understand the term. Of course that is exactly what happened on my second trip to the city of international cooperation and peace.

I was working in the river and hills when it came time to board a commercial airliner for my second trip to Geneva. The drive to the airport began the whole thing: metallic clouds of sharp red lay over misty grass for fifty miles, all outlined against the grays and wisps of a prairie morning. The sun broke as I turned off the superhighway, made my way through the sleeping town of North Platte, turned out over the river toward a toy airfield. The runway begins where the swallows fly. I have been to that airport many times, each time with families separating, sometimes parents, tanned cheeks but light foreheads, boarding a plane with propellers for connections to some far off land. That airport portends things that smack of tension. That whole town also portends tension because of a love/hate relationship that began seriously on the interstate not too far outside of town. I kissed my family good-bye, never knowing that one of them would be standing in that same airport with fierce tears of anger ten midnights later. My boots struck the aluminum ramp of a Frontier bird; the props were beginning to cough into life. We skimmed the top of cottonwoods at the end of the runway. Cowboy country slipped away beneath me. I closed my eyes.

Ten days later I boarded Air France in Geneva, headed for Montreal, Chicago, the state capital, then a prop job back into cowboy country. I was in a state of shock. The biologist within me cried out the frightening condition of the society in which I lived: our personal intellectual freedom is, *relatively,* so great that we do not recognize its position with respect to that of the rest of the world. As a biologist, I could only conclude that such freedom, as apart from the mainstream of world culture as some relict species might be from the evolutionary currents of its group, was perhaps in danger of extinction. The freedom, the rebellion against restrictions of the mind that I valued so highly, appeared so fragile compared to the forces of organization, regulation, policy, administration.

I had sat in a finely machined stainless steel chair and listened while a decision was made to send someone across an ocean to make sure a faculty member would actually be allowed to do the research he proposed. I had sat sipping Swiss café au lait, purchased from a machine for a half-franc piece, and listened while technicians' salaries were discussed. In the foreign nation in question, technicians could be paid with at least two kinds of money, both named the same word, but one bearing a certification. The certification allowed those technicians to shop at the same stores as the political leaders of the country. The other kind, without the certification, did not. Needless to say, the latter would be worth but a fraction of the former. I had sat, my mind wandering to an anticipated evening in the Pub Britannia, as the discussion turned to the availability of research chemicals and minor pieces of equipment I took for granted, or could obtain in a day. I had sat, amused at the upheavals in Swiss social structure that would occur when I—a consulting scientist—would actually sit down in plain view at an office typewriter, and begin work on a report while a request for funds from a scientist in one of the most populous nations on earth had to be returned because he had not obtained *national* approval to *ask,* yes, I said *ask,* for the money. I reviewed momentarily the events I mentioned earlier, the events I would experience in my own university *asking* for money. Now *getting* money is a different matter, but anybody can *ask* for money any time without anybody's permission. In fact, they often encourage it here! I remember distinctly it was at this point in those meetings that I decided to write this book.

Someone, somewhere, needs to be told what it is we possess out here. Someone somewhere needs to be told that the things we possess out here in cowboy country are really the things we *all* possess, everywhere, in this nation into which I was born. My concerns over the spreading of this message are, of course, wasted on one person, and that is my oldest daughter. She of all does not need to be told what true freedom of spirit it is we have as our birthright. It was also her tears that were flowing when I again stepped off the Frontier prop job at North Platte, home of Buffalo Bill, radio station KODY, airstrip starting where the swallows fly.

Boarding the plane that I was leaving was Congresswoman Virginia Smith, home from Washington for a few days and using the North Platte airport lobby as a forum for her views on the snail darter and Tellico Dam. Cindy had not shared her views. A confrontation between an elected representative to the Congress of the United States of America and a sixteen-year-old girl with her own opinions had occurred in a hurry. Sharp words were exchanged over the fate of small fish. The small fish did not even live in Virginia Smith's district. Tax money was involved. Values were discussed. I was in the stormy night skies of tornado-ridden Nebraska when all that happened, but I can hear so well that tone of voice, that choice of words, telling Virginia that money and politics are considerations that do not apply to the lives of darters. I can also hear Virginia's reply, for it was described to me in great detail through the tears. Cindy felt demeaned, patronized, by a politician's reply to her snail darter concerns. But then in all fairness, Cindy usually feels demeaned and patronized when the world does not bend to her will, and usually reacts accordingly.

Fifty miles further west, deeper into the dunes and prairies, we went that midnight back to our cabin in the hills, back to the marshes of Keith County. The confrontation at the airport, however, stayed in my thoughts. No secret police came banging on our cabin door in the nights of the weeks that followed. No one was arrested over that incident between a citizen and a politician. I was not harassed at my job, nor told to keep my daughter under tighter control. My salary was not changed to *un*certified dollars. Instead, Cindy vented her anger with a very pointed letter to Congresswoman Smith and Congresswoman Smith responded as some Congresspeople sometimes do to citizens. I am sure that if Virginia ever runs for anything more ambitious—Governor? (or is it Govern*ess?)* —she will never get Cindy's vote. But then that's America, is it not?

The decision to write this book had been made a few days before that midnight in the airport. The decision as to how to write it, and where to start, was made *that* night in North Platte. My heritage was shown to me in cowboy country. I made the decision right then that I would seek some *real* cowboys, and that decision was the one that led me to the Corfields and the Packards. They may not be the best examples, but there is little doubt that each of them, or at least the ones I talked to, expressed with total strength the river of freedom and independence that runs through this part of the world. I sat for a time in their homes, out on the range, with the memories of that midnight in the airport strong in my mind.

"Fifty miles further west . . . back to Keith County."

3

Home on the Range

MILLIE Corfield lives up south of Arthur in one of the more interesting homes in the Sandhills. I had stopped at the Corfield's three years earlier, asking permission to look in their well tank for snails. On that day three years ago I was pretty hesitant. The Corfield place did not advertise for visitors and the isolation of the Sandhills, an isolation that must be felt for it cannot be described, combined with my stupid snail mission, a fancy state-owned four-wheel-drive truck, and my city-slicker camera, gave me a sense of real physical danger. I could not have been more wrong about anything than I was about that sense of danger three years ago. That was my conclusion as I sat in Millie's living room and listened to her tell of their life on the prairies, watched Erle and Jim look at her as she talked, and finally turned to answer "No" to Erle's question:

"Ever hunt coyotes?"

It is obvious why I was sitting in that living room. The Corfield place was the origin of my strongest impressions of the isolation, the estrangement, the brazen self-reliance, that very pluck of the coyotes they hunt, the audacity to confront the elements that sweep down out of the north, which must characterize a family that would make a home on such range. My search for snail had, those earlier years, taken me throughout several quadrangles, over several hills that looked out only over more trackless hills as far as could be seen. I had stood so many times looking forward into that horizon, then turned to make sure I could see a landmark behind, could locate tire tracks from a different angle, before stepping high again into that cab for a windmill a map said was a mile away. But of all the places I'd seen, of all the permission I'd asked and not asked, of all my contact with the dry dunes and those that live upon them, none had been remembered with the force of the Corfield place. Back in Keith County, driven to write of intellectual freedom by those experiences in search of tropical health, there was, of course, only one place I could start to get the human answers I wanted. I'd like to think some of that Sandhills mettle has rubbed off after a few years of working out there. For in search of Sandhills mettle, I never considered any first stop other than the one that had already given me such apprehension. The Corfield place hadn't changed much, and neither had son Jim. He was still interested in the "sally-manders" in that well tank and could still speculate on how they got there.

Erle and Millie were out for groceries, so Jim and I talked for a while. The world goes away when you drive into the Corfield place, and it goes away even further when you sit down with Jim to talk about one-room schoolhouses. There was a time when I was out on the Sandhills right before sunrise, on a windless cloudless morning, and watched the special light diffuse out over the land. I could look east and see only one human mark upon that land: a one-room schoolhouse and a hint of a road leading to it. There was another time when a visitor from the East came to the Sandhills and we took him up south of Arthur to see some prairie. There was a one-room schoolhouse on the way to the prairie. We'd passed it dozens of times in the last three years, but the visitor wanted to stop. The schoolhouse was unlocked and we went in. The place was clean, immaculate, organized, with an irrigation company calendar, a curtained shelf with their little dried biology projects of a past year:

quart jars with seeds planted. On the shelf with the biology projects was the library: a stack of National Geographics from the Mesozoic. The whole thing was absolutely wrenching, and the strongest feelings of empathy, the strongest desire for success for these unknown children, welled up inside. The others felt the same, and the visitor from the East, a most sensitive man and teacher, expressed those feelings for us all. Back in the Corfield living room, you know what I had to ask Jim: what was it like going to one of those one-room schoolhouses? He answered without much hesitation:

"It was kind of boring."

I wondered then, and we wondered together, what kind of person would come to teach at one of those schools. That's when he told me all their names, every teacher he'd had down through the years, how long they'd stayed, which ones were the good ones, and how he got to school. There was an element of American grit in his descriptions of his teachers. A person might think, believing what we're often told is the American dream of higher, faster, more, that an ambitious person might not come to a one-room schoolhouse to teach. Furthermore, a teacher with ambitions might truly see one of those schoolhouses as the end of the world and wonder about the choice of a profession that started in a small white building with only the wind for company. At least those were my thoughts about teaching in one-room schools until Jim said something of the turnover rates of one-room schoolhouse teachers:

"Most of them stayed a year, then moved on. The better ones, they stayed about three years."

Right then I re-defined "ambitious" as applied to teachers.

We talked about social life, about girls, games, cars, jobs, his parents and what they could do in the future if they had to, or wanted to, or ever considered any life other than the one they were living. But always through such conversation ran a track of isolation, a track of sparseness, a track of closeness to the prairie. And at the end of that sheltered time, I came away feeling that maybe I'd met just another creature whose life was inextricably pounded into that econiche known as the Sandhills. He saw nothing strange, nothing unusual, about his life. Erle was the one who later would underscore the values that I could sense in Jim but couldn't quite get out, couldn't quite identify. Yes, Erle was the one, later, when he turned the questions back on me. It was the father, who told me through his questions what I was seeing in his son, when from the corner of a table he came out with:

"You live right *in* town, do you?"

I'd asked about the number of kids that go to those one-room schools and Jim had responded by telling me of the tragedy—the auto accident I'd read about at the other end of the state. That accident reduced their school's population to two, maybe only one. Later on, when Erle and Millie returned, we talked a little while of the administration of a one-room school with one student and one teacher. There were four in Jim's school, when he went there years ago, three girls and Jim. Years ago, Jim was in the eighth grade and the oldest girl was in the third grade. He told me that, then he made the comment about it being "kind of boring."

"It was kind of boring," he said. "We were supposed to play a certain game at recess, but I used to go play with the horses."

Jim works in Ogallala, too, as a "sanitation specialist."

"Not much to make a living on up here," he said. "People tell me I'm crazy to work in Ogallala, drive every day. Fifty miles an hour saves a lot of gas, but still, the gas prices has really hurt, and that job's not really what it ought to be." For the historical record, this conversation took place on July 2, 1979. Gas was pumped in Ogallala at under 90¢ a gallon in July, 1979. The sentence was written on January 13, 1980, and for the historical record, gas was then pumped for $1.07 a gallon. That's an average, estimated, for leaded regular. I'm sure Jim's Ford pickup, parked outside their home on the range, burned regular. None of those kinds of pickups ever burned premium.

"I've been to Cheyenne, Sioux City, Texas, and Chickasha, Oklahoma. Not much to make a living on up here, but if I had my choice, I'd make enough money to live right here and be the happiest man in the world. The folks own one section. Went to college one semester, but even if I'd had four years of business, I'd still have come back to live right here. Oh, folks could sell this section, move into town, buy a house, make nine, ten thousand a year off inflation and interest on what they'd get for the land. They'd never do that. We're not money hungry, just prefer the horses and cattle. They leased the place in 1948; bought it in 1962." Jim looked around his living room, then he said "Nobody ever so much as give Dad a nickel."

"I'd like to get a picture, Jim."

."Let me get my hat and boots; wouldn't really be a picture of me without my hat and boots." Nor without his can of Skoal, although he didn't say that. Back out in the Sandhills brightness, a car moved through the distance along a road I knew was there but couldn't see. "That's them comin' now." I asked one last question before Erle and Millie pulled into the yard. It was a question I had to ask, standing in the July heat.

"What's it like out here in the winter, Jim? What was last winter like?"

He reacted almost as if I'd kicked him in the foot.

"It's a bitch!"

I took a couple of pictures of Jim in front of their home on the range as Millie and Erle drove in. I introduced myself.

"This fella's a writer," said Jim. "Wants to talk to us about what it's like living out here."

Erle didn't say much, shook hands, then kind of turned away watching me out of the side of his eye. His name was across the back of his belt, and as he turned away I memorized its spelling. Millie had other concerns, I suppose from having survived in the Sandhills.

"Let's get these groceries inside," she said. Inside she offered iced tea, sized me up, and was way ahead of my question. "If I had it to do all over again, I wouldn't do a single thing different," she said.

Erle smiled from the corner of the kitchen table and rolled a cigarette, the first of several he went through while we talked, but it was Millie who was ready to tell of life in a home on the range: "When I was younger I worked in Omaha for about six months then came out here for a vacation. I had every intention of going back, but stayed. I would never do a single thing differently. We are free, we have our health and our land, our kids are on their own, we've done it on our own." My pride at being a part of this nation where a person can say those things almost brought those emotional tears, and as I write those emotional tears come back. Yes, Millie, you are free, you have a freedom that so few others know, that three billion people do not even understand, and best of all, *you* know it! Awareness, is that not the essence of human, an awareness of self, an awareness of life, of death, of joy and pain, of status? An awareness of *freedom,* of *independence,* is that not the ultimate essence also of being a human, of realizing all the potential contained in that word "human"? Yes, it is. Of course Erle knows it too, and that one section of Sandhills dunes that he and Millie bought in 1962 has worked wonders on his perspective.

"What's California got that we don't?" he said in a voice that could have belonged to some Senator—deep, polished, confident and secure, at the corner of his own kitchen table, rolling his own cigarette. "What's California got that we don't?"

Nothing, Erle, California's got nothing that you don't have. You own your own land and you are free, an independent man. You fit my definition of a cowboy. There is nothing California can add to that.

"I'm a teacher," said Millie, "in Arthur. I teach six, seventh, and eighth grades." That brought the discussion back to one-room schools, of children who go to them, of their ambitions, their performances, and of the teachers who teach in them. Millie was one of those teachers once. "They rotate teachers in and out of those schools. You teach all the subjects, all in one room, but sometimes the art and music teachers come in just for that." We talked for a while about her students, the ones in Arthur, then in general about school kids in Arthur. "Math is easiest to teach," she said. "I like it the best and I do the best job at it."

"Where'd you get your college education, Millie, your teaching certificate?"

"I never finished high school," said Millie Corfield. "It was the Depression. I couldn't finish high school."

Seems Erle had found a matchbook not too long after they were married; probably picked it up somewhere to fire up one of those rolled cigarettes, rolled while he contemplated some piece of land or some coyote out ghosting across it. The matchbook had an advertisement: HIGHSCHOOL BY CORRESPONDENCE. Millie took the matchbook up to her friend, Annie Mae at the courthouse, sought another opinion on whether the advertisement was legitimate. Annie Mae wasn't too impressed with the matchbook. Instead, she told Millie she ought to take the GED exams, General Educational Development exams. If you passed, you got a high school equivalency certificate. They threw the matchbook away; wasn't much question in *her* mind that she'd acquired the General Educational Development equivalent of a high school education. She passed, went to Kearney State, got her twelve hours and certification, and came back to the Sandhills. Kearney is still Nebraska, it's still on the border of 20,000 square miles of Sandhills, but it's not actually *in* the dunes. Millie came back west with her certification and now she teaches middle grades in Arthur. Math—that's her favorite, the one she feels she does the best. They still print those matchbooks—HIGHSCHOOL BY CORRESPONDENCE. The last one I saw was in the Denver airport. It reminded me, as they always will, of how Millie Corfield became a math teacher in Arthur.

My notes, my scribbled ideas, hasty observations and things I must remember, tend to be kept on any handy scrap of paper. I clean out my pockets every once and a while and put those notes in my briefcase. Then about every six months I clean out my briefcase and put those notes in a file. You wouldn't believe that file. Actually there are several such files. One of these days I'll take all those files and put them together in a file cabinet, and on and on. So a couple of years ago I bought this little red vinyl covered notebook with the seal of the University of Nebraska on the front. It was quite a joke when I first bought it. There wasn't a single thing on any of the pages, and I used to tell people it contained all there was of significance about the state of Nebraska. That's no joke any more; my notes taken that day on the Corfield place are in that red notebook, but I'm wondering why I remember so much more about that day than is written in my notes. There is a message to me in the medium of those pages, and the message is this: where those notes dribble out, where there are only a couple of sentences to remind me of an hour's conversation, those are the times the conversation hypnotized me so I could no longer concentrate on the notes.

I wish I could convey Erle's voice, his way of telling about hunting, but not really telling at all, just having that certain look on his face when the subject came up. You could tell by that look something of the hunting of *Canis latrans* upon the dunes of western Nebraska—or at least you could *imagine* it, but your image might not even be close to reality. Then we talked about town, about Ogallala. Too much of an urban influence to move their kids into Ogallala, said Millie, and Erle agreed. That's when he asked me if I lived "right *in* town, right *in* Lincoln?" I don't think he'd ever been that close to a person who lived right *in* a town of 160,000. I said my good-byes and they said their come-on-backs. Even then I wondered why I'd come, what right I had to simply drive up to the Corfield place, ask them about their lives, then write it in some kind of a book. Maybe it's not a right, but a responsibility instead. Maybe I have this responsibility to put down for someone somewhere my impressions of a couple of hour's conversation with people who *know* they are free, independent, have their own land, and who refused to move into town when their kids were small. Maybe that's a responsibility we all have.

If you don't remember what late 1979, early 1980, were like, then I can remind you. Americans were being held hostage in the embassy in Iran. Russia invaded neighboring Afghanistan. OPEC's biggest internal squabble was over how much to raise crude oil prices, how much the United States could afford, the upper limit that would sustain the apparent American addiction to energy but not generate true backlash from the American citizens. The concept of a nation so free that its own citizens, its own press, could express their opinions without fear of reprisal, was a concept so foreign to the world of 1980 that it resulted in ridicule of my land. Maybe this is only my outlaw approach to life coming to the surface, but I no longer feel any qualms about saying the things said all through these pages: there are, out in Arthur County, Nebraska, a lot of cowboys who *know* they are independent, free, a lot of people whose *awareness* of their freedom is a lesson in higher civilization itself. I left Millie, Erle, and Jim, turning out onto the blacktop west, back toward the highway south of Arthur.

A mile down that road toward highway 61 is another gravel drive, this one leading up to the Packard Ranch house. On that same trip three years ago after snail, that same trip where I'd first met the Corfields—without really meeting them—I'd also tried the Packard Ranch. No one had been home, then, but the dogs were home. You should know there's a reason strangers are wary of ranch dogs. We'd been able to raise no one at the Packard Ranch house three years ago, and I was very disappointed. There were a number of wells on their property that I wanted to check for snails, and that fact alone, once you've looked at a Sandhills map, tells you something about the Packard Ranch. I remember thinking at the time that by getting permission from the Packards, we could visit four or five, maybe ten wells. Sandhills water wells are surprisingly evenly distributed. Throughout northern Keith and southern Arthur Counties, Nebraska, there is an average of 1.2 water wells per square mile.

Shirley Packard was making salad, Gerry was gone somewhere, and one of the Packard girls was mowing the lawn, when I pulled into that gravel drive to talk about a home on the range. The Packard home on the range is fairly well known outside of cowboy country, and in fact the words "Packard Ranch" are on file with the federal government: there is a U.S. Geological Survey topographic map, 1:24,000, rolled in my map file, marked with numbers on water wells, worn from being out on the dunes. That map is officially titled the "Packard Ranch Quadrangle." The Packard Ranch quadrangle is bordered by others with equally romantic names: Bear Hill, Glinn Ranch, Martin, Lemoyne, Spotted Horse Valley, each name identifying the major geographical or sociological feature of that section of America. There is nothing like a U.S. Geological Survey topographic map to give you the wanderlust; mostly even the name of some quadrangle will do it. Often there's a special feature of some kind that you can see on a 1:24,000 map—a geographical feature, a long canyon, a steep canyon, maybe a marsh. A 1:24,000 map is large enough to show even the numbers and placements of buildings. On the Packard Ranch quadrangle you can see the Packard Ranch buildings, their relative sizes, their placements, and you can see another special feature of the Packard Ranch quadrangle. A small airplane indicates an airstrip. You can also see the wind sock from that blacktop road east of highway 61.

"Gerald H.—that's Gerry's dad—he flies," said Shirley Packard as she handed me her copy of *Keith County Journal* to autograph.

Shirley Packard, like Jim Corfield, had some comments about the winter past. As I type these words it is also winter, mid-January, and ten after five in the morning. There is a gentle rain falling, the first in many weeks. It was in the fifties yesterday, and the day before that, and even back through the weekend, when I spent a day of the prairie winter doing what people normally do in April: soccer with John III, a minor tune-up for a balky vehicle, an hour sitting on the front step with a beer watching America go by. For some stupid reason I was watching television late last evening. There was a smell of spring, a false smell, of course, in the night air, and the weather girl was on. Television is not one of my major forms of entertainment, but when it's spring on the prairies in January, it sometimes makes you curious enough to watch the late evening weather. A year ago almost to the day, the high was 0°F.; yes, zero was the high. Now I understand there are colder places in the world. But I also understand that when the high is zero and the wind comes ripping down over the dunes from North Dakota, then it's time to take shelter in a home on the range. My notebook doesn't indicate what my question was; somehow I assumed I'd remember it. But it must have concerned winter, since my notebook does indicate Mrs. Packard's answer:

"After *last* winter that's a hard question to ask. Last winter was tough." The question must have had something to do with the nature of Packard Ranch operations. I must have asked what their major business operation was, which of their activities provided most of their income, occupied most of their time, and how they did those things in the winter.

"Gerry's grandparents homesteaded here in the early 1900s," she said. "There are eighteen sections total. We have cattle, of course, and hay, corn, rye, alfalfa, and sorghum. We do all our own mechanical work. The horses are mainly for personal use. We've put in some center pivots, the first one about ten years ago, primarily because of the shortage of hay in the dry years. But we've also had to diversify because of the cattle prices. Corn made more than cattle last year." We talked on for a while about her kids and their plans, about Gerry's stint as a Marine in San Diego, Florida, Texas, about *Keith County Journal,* until the talk inevitably—because I'm a teacher, probably—got around to the one-room schoolhouse, especially a certain one on the Packard Ranch quadrangle.

"Petersen's hired man's little girl will be the only student in there until Christmas. Petersen's hired man and his wife, they'd be the ones to ask about one-room schools right now. Their little girl will be the only one in there unless someone else moves in. Another kid is supposed to start kindergarten there after Christmas. *That* school is a result of the consolidation." She paused as I wrote the words faithfully in the red vinyl notebook containing all there is of significance about the state of Nebraska. Those were evidently her exact words, for they are written so legibly: "That school is the result of the consolidation."

"All three of ours went to the one-room school. Both Gerry and his brother went there. There's nothing wrong with a one-room school; it's great if you have enough kids."

"How many is enough?"

"Enough kids is as many as it takes to instill some competition."

"Do you remember their teachers?"

"All three had Millie Corfield as a teacher."

I finished with the question I was after that day, July 2, 1979, two days before the 203rd anniversary of the signing of the American Declaration of Independence. And Shirley Packard answered with the only answer I've gotten from anybody in that country, the same answer that has come in form after form, from person after person, in medium after medium, in place after place, wherever boots slip a little bit in the sand and wherever the needlegrass tries to work its way into a pair of jeans.

"Some folks stay for sentimental reasons," she said, "and a lot of them won't admit it, but they're staying for independence, for dependable neighbors, for the freedom of owning their own place, being their own boss, that's why they stay, for independence."

The Nebraska Sandhills, Cowboy Country, U.S.A., covers approximately twenty thousand square miles. The Corfields own one of these. The Packards own eighteen. There are those who live there in the midst of the dunes and own none. There are others who cannot see the ends of their properties. Out across those twenty thousand square miles people have chosen not to leave. They'll tell you where a burrowing owl nest is, for they've seen it from horseback. They'll tell you about the deer and they'll tell you what's in their well tanks. They leave their foreman in charge while they go off to rodeo. They drive pickups, put that pinch of Skoal between their teeth and gums, and stick a rifle up in the rack across the back window. They fight the north wind that roars down upon them in the dark and when it's all over they're out there in the late spring a year older. There are more kangaroo rats, voles, wild roses, earless lizards, lark buntings, deer flies, and swallows out there in those twenty thousand square miles than they can imagine, much less count. They're uncomfortable in town; at least they *look* uncomfortable in town. Some folks would say they live in the wilderness, out there in the middle of all those dunes.

But other folks know better, know things about wilderness that not all of us know. These other folks know that when you can look around at the dunes, say you own some, say that nobody so much as ever give you a nickel, say that there are more kangaroo rats, voles, lark buntings, wild roses, and earless lizards than you can imagine, that you've made a home on the range with them, then these other folks know you are free and independent. And these other folks know *you know* you are free and independent. Some city people might call you a cowboy, then, when it's obvious you know you are free and independent.

The Nebraska Sandhills is a recreation area. Hunters come, and fishermen, all with much equipment, boats and the like. Game and Parks tries to supervise Big Mac, the biggest lake in cowboy country. The Fourth of July weekend is big business around Big Mac. I came out of the Sandhills that weekend in my white Merc with the black stripe, tape deck blaring Waylon Jennings, red vinyl notebook straining with the individuality of the Corfields and Packards, into the rush of holiday sunburn, campers, ice, fish, boats, motorcycles, sand, and blue water twenty-two miles to Lewellen. I stopped at Sportsman's Complex for a beer. Some guy was buying bait.

"Got any shiners?"

Right away that reminded me of a kid named Mark Safarik. "Shiners" had taken on a new meaning over the past few weeks. The new meaning reminded me that the unyielding individuality I'd found that afternoon up south of Arthur could be found in other places, places such as the brains of students. "Shiners" was one of four existing types of fish, said Mark, and off went his brain into a flight of fancy. He also wrote songs; some of the better ones were about worms. You must know that when a person writes songs about worms and reduces the world to four categories of fish, then there is in that person a creative individuality worthy of study. Maybe that person lives in his own home on the range, his own home of his own creation out on that range of unfettered thought that stretches to infinity, that rises in marching dunes through history. And maybe when that person has homesteaded his own home on such a range, then that person is also a cowboy of sorts.

4

Game Fish, Trash Fish, Chubs, and Shiners

THE title of this chapter was first told to me by a young man named Mark Safarik, although I heard later that another young man, Jim Krupa, had actually synthesized the phrase "game fish, trash fish, chubs, and shiners." Krupa and Safarik were not necessarily good friends, but they were certainly kindred souls. Krupa was by far the more practical, while Safarik was simply a dreamer, lyricist, and songwriter. Unfortunately you are never likely to meet either of these young men again unless some day I decide to write an entire book on one of them. Either could support such an endeavor. On the other hand, if you are a member of the select group that had the opportunity to share Keith County with Krupa and Safarik, then you will instantly recognize small bits and pieces of them throughout the rest of this book. I think, however, that I've dismembered them enough to avoid positive identification in most cases! Krupa evidently synthesized the phrase "game fish, trash fish, chubs, and shiners" to describe the practicality with which Game and Parks viewed piscine resources. Safarik, on the other hand, immediately seized upon the phrase and extended it through his soaring creative originality into what I've since come to call Safarik's Taxonomy. This classification scheme, of course, has nothing to do with fish. Safarik's Taxonomy applies to people.

I knew a fisherman once, out in cowboy country, who might have been the best fisherman of all. In a moment I will tell you his secret rig for catching trout. He certainly hauled in the trout, but then he hauled in a young lady with the same rig. She was the essence of athletics. She refused to be beaten at the games she chose to play. He also refused to be beaten at the game he chose to play, that of the fisherman. Not many people who call themselves fishermen take such pleasure in catching a chub a foot long. What most people buy for bait, this fisherman sought out as an unusual eating experience, some minnow grown grotesque in an isolated pool up in the headwaters of some cowboy country stream, some relative of the carp, some glob of raw sucker eggs, all prepared a certain way, all in addition to the good stuff: trout, walleye, channel cat. A fisherman who eats the fish that most so-called fishermen use for bait is a fisherman who is eating fish other fish choose to eat, given an opportunity. That had to make him a game fish. He caught game fish with his rig—her included—so that makes her a game fish, too, now doesn't it? Before you get all excited about this story of man catching girl with spinning rod, you should know that he was married, in his mid-thirties, and she was thirteen.

They fished what was called a canal, although it was really just part of the river. Its only function was to keep water flowing through a lake, which was also part of the river. He used to go down to the canal every morning early and throw out his fly, at the end of a 7-foot 4-pound leader, with his spinning rod and a water-filled float for weight. He did the same thing every evening. And every morning and every evening he would come back with trout. And every morning and every evening boats would go down the canal, laden with sunburned people out of North Platte, and every morning and every evening the boats would return, and every morning and every evening we would call out to those boats, asking whether they'd "had any luck" and every morning and every evening the answer would be the same: "Naw."

A good many other fishermen tried the canal. They tried their own rigs, their own ideas of what it should take to catch a game fish, but somehow these rigs never did work with the regular success of *the* fisherman's spinning rod, water-filled float, and nondescript brown fly. The girl watched all this for a few days with a calculating eye. If there was one thing she could spot at the horizon it was a contest, no matter how subtle, no matter how obvious, even between adults, no matter what the medium. This best of fishermen was small of stature, like herself, and the others often called him "Captain Minnow" behind his back. This best of fishermen was gentle, like herself, and he made friends easily, like she did. But of all the ways in which they were similar, they were most similar in one: a certain kind of inner smile crept across their faces when they did things others could not do. Like catch trout. Or catch baseballs hit a certain way, hard.

Her father had sensed that perhaps this was his family's last summer in cowboy country and had brought along a spinning rod, one given years past to another daughter by a grandfather now gone to his last fishin' hole. Her father, sadly but with some cunning, had tallied the things he should perhaps try *this* summer—sensing it might be his last for a while—things he had put off since childhood, things like fishing. But he was ill at ease with the newfangled equipment that others seemed to handle with such familiarity. She had never handled a fishing rod of any kind, period, when with the most gentle, but most firm of gestures, the softest of touches of girl becoming woman, she lifted the rod from her father's hands and headed for her own place on the canal. It was evening when all this happened, and her father watched for a time from the bridge over the canal as she wended her way through the weeds, casting out across the canal, having never casted anything anywhere before this evening in the deepening twilight. The mosquitoes came and he called to her that it was time to go. But she called back, "Just a minute!" The mosquitoes came again, and again he called to her, but again she called back, "Just a minute, just one more try!" The mosquitoes came again, this time with seriousness, and he called to her again, but she didn't ask for one more try. She didn't have to. He should have known it would happen this way.

"Dad! I've got one!" It all sounded so simple but there is no way to convey the tone of that voice. Imagine a violin in the hands of some special person; then you can imagine the tone in that voice from down the canal out into the dark. Pictures were in order, back up in the building where we worked in the lights. Everyone was proud. She had caught a fish, a "real game fish" as some would say. But Captain Minnow smiled his certain smile and she smiled her certain smile, and he knew *he* had caught a game fish. For she would fish again and again, and he knew he had caught her, bent her desires to a human activity he placed at the highest pinnacle of value, simply by being best at that activity. And her father looked later at those pictures of her first fish, a trout, and decided it was not nearly as large a trout as he had remembered it being that evening weeks before. And his eyes turned toward his daughter in that picture and the words "real game fish" came again into his mind and he did not have to be told which of the creatures in that picture was the real game fish.

Game fish are game fish only in relation to trash fish. If there were no species called "game fish" then the species we call trash fish would probably be fair game. Captain Minnow said many times what a delicate taste trash fish often have, which probably means that for a true game fish, "fair game" has a much broader meaning than perhaps we suspect. I did see him slit open the belly of a white sucker, *Catostomus commersoni,* once, and eat a handful of eggs right out of the fish, raw, with a little salt, all the time making some comment about caviar. I tried that caviar then, casting an eye toward Captain Minnow and deciding that he probably behaved that way all the time, that he was in exceptionally good health, so sucker eggs right out of the fish right out of the lake couldn't be fatal. They weren't. But now that I've tasted sucker caviar, I'll probably choose beer next time. The white sucker has never been known as a game fish out here in cowboy country. No, around here the white sucker is generally regarded as trash.

There are some interesting facts to be stated about the white sucker. First, the white sucker loves deep swiftly running water, usually cold. Right away we see a paradox: the word "sucker" conjures up some mud-eating sloth out of dirtier parts of the water. But the real sucker is most often seined in the places we associate with game fish. I wonder if the sucker would be trash if it stayed away from the places game fish are supposed to be. Secondly, the white sucker is a delicate fish, and large ones do not survive even the slightest attempt at domestication. Evidently they require more oxygen than a bluegill.

Suckers also have a plaintive look to their faces, enhanced, perhaps, by an enlarged sensitive area beneath their chins, their rounded (balding?) foreheads, and their, well, sucker eyes. And they're hard to identify as immatures. I have this suspicion that we could all walk up and down the banks of fishin' holes everywhere with a certain kind of specimen, offering it to old folks, kids—fishing, almost, for an answer—and on every bank we would find person after person who would correctly say, "That's a little bass." But this is only another form of fishing, this casting out into intellectual waters to see what's there in the way of true understanding. So what if we used another bait, a trash fish? How many would say, "That's a sucker fry," or "That's a first-year suckermouth minnow"? Not many. But there would be one, somewhere, who would take the thing in the most gentle of manners, would cup his or her hand and dip it into the water to moisten those gasping gills, would spread the dorsal fin, would smile and say, "That's a little *commersoni.*" This game fish would hand you back the sucker and watch what you did with it—whether you threw it back up into the weeds or whether you tossed it lightly into the water. And you would know you'd caught a game fish by the look on the face if you tossed the sucker into the weeds. And this game fish would know by that same action that he'd struck at trash.

We live today in a world of heightened sensitivities to the wilderness that supports us all. You may not believe that but I see it every day in the students who populate my classes. I sincerely believe there is a movement afoot to change the way we deal with our dwindling resources, for I have had to make some major adjustments in my own use of game fish, trash fish, chubs, and shiners, in order to accommodate that changed mood and still do what I'm paid to do: teach biology. For example, one of my favorite biology lab exercises involves the white sucker. This exercise is a favorite for two reasons: first, it always works; and second, no one loves the white sucker, especially Game and Parks. This exercise involves a worm that lives within the white sucker, a worm called *Monobothrium,* as well as some fairly sophisticated mathematics. Most students, especially pre-meds, greatly appreciate this class exercise because it always works, but in addition, I sometimes think, because of the white sucker. Generally there is no emotional response to the killing of white suckers. Generally. I really didn't count on an emotional response the last time we gill netted white suckers. But then I guess I really didn't know Kathy very well.

The idea for gill netting suckers came from Game and Parks. Game and Parks gill nets all the time as part of their research, and in certain lakes most of their haul consists of white suckers. Trash fish. Now it's a well-known fact that white suckers have these worms, *Monobothrium,* and that they get these worms from eating other worms. So all you have to do is stand around and watch Game and Parks haul in white suckers and right away you have this great idea about the mathematics of sucker worms. Here's how it works:

The distribution of *Monobothrium* among suckers can be described by a mathematical equation known widely as "the negative binomial." If you draw a graph upon which the number of suckers is on the vertical axis and the number of worms/fish are along the horizontal axis, then a bar graph with the general shape of "the negative binomial" almost always results. What is of such cosmic significance in this matter of worms, however, is that such a graph comes very close to describing almost every distribution of one kind of animal that is symbiotic with another. This situation is especially true out in nature. When it is true, and the symbiont is a parasitic worm, then you know that the *majority of the hosts are not infected and the majority of the worms are found in the minority of the hosts.* We seem to be straying from the subject of gill nets, Kathy, Game and Parks, heightened environmental consciousness, and so-called trash fish, but not in my mind. In my mind they are all meshed together with the negative binomial. It all had to do with this rather routine lab exercise, the one that always worked.

I was returning from Switzerland, due back in cowboy country at midnight, and due back in class at eight the next morning. In anticipation of eight o'clock class, I'd asked my assistant to set the gill net. There was no way we could fail. We always got fish. They always had worms. The worms were always distributed according to the negative binomial. The calculation of theoretical curves describing the life of this worm in nature was always a challenging, interesting, satisfying exercise for my classes. And for me. But I hadn't counted on Kathy.

You have to operate a gill net in order to understand one. They are pretty simple devices, ingenious in design, effective beyond words. As with most nets, they have a float line and a lead line, and in the case of ours, those lines are very long. One end is fastened to an immovable object, the net is strung out through the water, a gigantic weight is attached to the free lead line end, and a large float is added as a marker. What lies between is a three-hundred-foot-long, eight-foot-deep curtain of mesh. Into the mesh swim the fish, through the mesh swim the front ends of the fish, and it's all over. The gill opercula flair and become entangled in the net. No fish, once caught, ever got out of any gill net. It takes a human hand to free a fish caught in such a net, and even then, such fish are often reluctant to swim off. The gill net seems to sap the strength of will. Suckers released from a gill net often make the motions of swimming away into the freedom they once knew, but the defiance of will just does not seem to be there.

Our lake, the aforementioned one with the canal, is full of white suckers. So on the morning after my return from Switzerland, I found myself bouncing over ragged waves beneath a threatening sky to the far corners of water where lay, like a curtain of cold, three hundred feet of our gill net in the water. The rattling obscenity of outboard engine was punctured by the slam of those gray waves on the aluminum hull, beneath where I sat unshaven, a cultural hangover temporarily resisting the curative effects of sand, hard grass, and Chevy pickups. The float appeared, bobbing and disappearing, coming closer at some unjudgeable speed, until with a swoop, watch shoved high up on my arm, the strangling cough of that cut engine echoing off shoreline willows, I hooked up the empty plastic gallon milk bottle, jerked it into the boat with a spray of lake water, weeds, algae, mud, and pure organic smell, and began pulling in white suckers. I pulled and the net gathered in loops in the bottom of that boat, suckers flipped, and my driver shook his head and fumbled for a cigarette. The suckers came, and I pulled more, and came more net, more algae, more suckers, and it all went on like some surreal dream of fish and weeds and water and smell until the little boat struggled with the weight of suckers, hundreds of suckers, hundreds of pounds of trash fish. That part of the morning ended with a vision of fifteen or twenty strong people carrying large buckets of suckers up into a building, killing suckers, cutting up suckers, counting worms, eating caviar, and when it was all over there was an accumulation of data enough to write a doctoral dissertation, but still there were suckers in buckets. The strong people trudged back down the hill with buckets of suckers and sucker remains to bury it all.

It was only later, later after the mathematics, after washtubs of suckers were buried under the sand, after some days had passed, that Kathy had an opportunity to say something about trash fish. I don't remember the circumstances, maybe we were walking the fields, maybe sitting with a crowd in some dank bar, but I do remember the subject of sucker math came up for discussion. She was a laboratory person come West to learn about nature. Evidently to a laboratory person a white sucker is not much different from other fish. There was a lingering resistance on her face, not a sadness necessarily, but more a resistance to the mood of success we felt over the sucker math exercise. She obviously did not think it had worked all that well.

"We should have done them all," she said. "We should have gone through all those suckers."

We'd buried maybe a hundred without looking for worms. There were just too many suckers, evidence of the gill net's efficiency. We'd spent enough time cutting up suckers and counting worms. That was my conclusion. I'd never thought much about those ones we didn't use. After all, the exercise had worked perfectly. It always does. But she had thought much about them and the fact that they were captured, buried, no use having been made of them. It had obviously preyed on her mind for days, since we'd strung that curtain of death through the weedy waters of Keystone Lake. Somewhere out there are people who would heartily endorse what we did: gill net up a bunch of suckers and bury them all, including the ones we didn't use. "Couple hundred less trash fish," they would say. But Kathy is not one of those people. We'd taken more than we'd needed and committed the double sin of not using all we'd taken. Only her inability to see the difference between game fish and trash fish, a difference built upon human values, a difference superimposed on whatever natural differences might exist between the two, only that inability allowed her to see what we had actually done.

"We should have done them all," she said again. I had the feeling she felt we could have avoided at least one of the sins.

I've thought a lot about the gill net since that experience. Kathy is gone to Texas with Jim and the net is coiled in a tub stored in an attic out in the west. But there are days when I see gill nets everywhere, hanging in wrinkled sheets through the weeds of a shallow existence. And on those days I see what some would call trash fish winding their ways through that existence only to get caught up in a tangled mess that can only spell the end. I see what some might think are game fish also caught in those nets, the force of a predatory decision carrying them into circumstances beyond their opercula, entangling them in circumstances more confining, more fateful, than at first envisioned. You must know it takes a human hand to free a fish from a gill net. The design was conceived by humans, put into reality by humans, that curtain must be set by humans, and a fish, once caught, can only be freed by the multiple options of human hands and fingers. And of course that also goes for all those other gill nets drifting, hanging, through the weeds of a shallow existence, and for the fish that get caught up in them. Be they game fish or what some would call trash fish, it still takes a human hand to free them.

But the white sucker never seems to leave the net unscathed. A freed sucker never seems to have the driving will, at least at first. There is some fish memory, no doubt, that is unable to casually shake off the experience of entangling circumstances. Every time I see a sucker so freed from a net, slowly shaking off the effects of entanglement, I conclude that the sucker will once again acquire all the strength allowed *Catostomus commersoni,* but that that strength will not return until the sucker is some place out in the lake. Keith County, I conclude, there is a Keith County for white suckers out in the middle of our lake, and that is where those fish go to regain their strength, to shake off the effects of their entanglements!

The creek chub, *Semotilus atromaculatus,* is one of your more interesting fish. First of all, it's ubiquitous. I've never caught a chub and even considered that maybe it didn't *belong* where it was caught. These observations inspire some suspicions in our minds about an image of an ecological niche for a species. Sure, the chub has one, we say, and maybe there are those of you out there who know infinitely more than I about the creek chub *S. atromaculatus* and who are able to delineate that *one* ecological niche of this species. I would be interested to hear your delineation of that niche!

I wonder if it's the same niche wherein lives the English sparrow, the starling, the cockroach, the house mouse, the Norway rat, *Paramecium aurelia,* and other no-goods and blots-on-the-town and bums.

Ecologists nowadays speak of niches in terms of resources. They call time a resource, be it time of day or time of year, and they draw a time line, adding to that line a bracket indicating that portion of time occupied by Species X. A line is a one-dimensional structure. Then ecologists will call temperature a resource and draw a line that depicts a range of temperatures from absolute zero to the temperature of the sun's core. Then they'll add to that temperature line a bracket indicating the portion of temperature occupied by Species X. Then they'll call space a resource, and measure it somehow, then draw a space line and add to that their bracket indicating the portion of space occupied by Species X. Then they might measure the concentration of salt in the environment, finding places where the purest of water, distilled by some natural forces, filtered by some chance resin, is totally free of salt, but also finding places such as the ocean, which if it has one familiar characteristic, it is salt. Then they would call salt a resource, draw a salt concentration line and add that bracket indicating the limits of salt concentration within which Species X can live. One line is one dimension. Two dimensional objects and figures have two lines (Intersecting lines define a plane). Three dimensional objects have three lines (The formula for the volume of a cube with a single side *a* is: $V = a^3$). And since this is the space age, and we are all familiar with the problem of relative time, it is not difficult to imagine a fourth dimension, especially when that dimension is time. But anyone who has gone to the field after creek chubs knows full well that four dimensions, space, time, temperature, and salt, are not enough to even begin to describe the place where lives *Semotilus atromaculatus.*

There is the matter of the speed of the water, and of the depth of the water. There is the matter of the size of the chub, or its age, which must be related to size, for a niche must include the places they all, great and small, live. There is the matter of food, and a dimensional line for food could be drawn in any number of ways: the species of food for chubs could be graded and aligned according to some scale, perhaps according to caloric value per unit of weight, perhaps according to nitrogen content per unit weight, perhaps according to the amount of energy a chub had to expend to get that standardized unit of food, and you must know that each of these ways of delineating food could easily be another dimension of the creek chub's niche. So now we're up to nine dimensions of a creek chub's place to live. But there is little doubt a person could go tromping off across this country, seine dripping slung over shoulder, jeans wet to the thighs, after creek chubs and find not only chubs, but also the fact that even nine dimensions don't begin to describe the life of *S. atromaculatus*. Before it was over there'd be ninety dimensions, or nine hundred or nine thousand, and that person would then be standing in some hall filled with computer terminals asking a *machine* to spout forth the life of the creek chub according to the equations for dealing with n-dimensional hypervolumes.

A biologist would have to call the chub "successful" in the same sense as sparrows, starlings, and humans are successful. There are lots of them and they occur across broad geographical areas. They've arrived at those places by exploration, by exploitation of the freedom to gamble their lives away upon the chance of finding a bountiful place to live. But I find myself wondering who has been replaced in those places now filled with creek chubs. Which species became extinct when *Semotilus atromaculatus* poked its little nose into some stream and thus entered that game called "competitive exclusion"? I find myself wondering about the long term effects of success, at least as chubs, sparrows, and humans manifest it. If a species, simply by being successful, can exclude other species from one of the dimensions of its life, then is that species reducing the richness of its own environment? Might not the species replaced have had something to offer, some unusual habit or morphology, to which some chub down the line might respond in some way?

Now this next question is going too far, but I'll ask it anyway:

Was there ever a chub that resisted its own species' willingness to play competitive exclusion? Was there ever a chub that deplored the extinction of some less successful minnow? All chubs deplete their environments, if only of the proteins, lipids, and carbohydrates they consume. But did this one chub also see in the game of exclusion an environmental depletion of a more serious nature, of a resource that could not so easily be replaced as temperature and salt? And did this resource get entered into some ecologist's computer as the nine hundred first resource: The cultural richness contributed by some species other than a chub? Some dilemma, it is, to choose between one's own immediate success and the continued existence of a species one would normally exclude from one's own environment!

But of course the chub has no choice; its genes drive it to exclude that other species. The freedom to move through cowboy country creeks, a freedom that in analogy we so admire, protect so fiercely, is a gift of genes, chub genes, but that freedom is a double-edged sword. Or rather, that freedom *would* be a double-edged sword if there existed that one chub that saw the value in some species other than chub or chub food. I cannot help but go back now to that list of species whose ecological niches are indeed mammoth hypervolumes, that list of most successful species: the creek chub, the English sparrow, the starling, the cockroach, the house mouse, the Norway rat, *Paramecium aurelia,* and, of course, *Homo sapiens.* None ever lacked courage when it came to exploration, to exploitation of an open niche, to the exclusion of less competitive species. But which, I must ask, upon that list has at least the potential to forcibly override its genes' commands? Which upon that list has the genetic ability to choose, to opt *for* a hypervolume that contains the snail darter, the blue morpho, the buff-breasted sandpiper, the California condor, the red wolf? And does that choice not also involve, even require, the same courage it takes to be a cowboy anywhere?

Of all the fish in cowboy country, it is the shiner that is most used, but least known. Of course there is no *the* shiner. There are several species called "shiners" as well as some species whose common names include the word "shiner." Red shiner, golden shiner, shiners swimming in the Platte, shiners swimming in a bait tank, shiners hanging on a hook, plastic and metal shiners hanging on the racks of tackle shops, every one a shiner, every one destined to fulfill the role given it by a human: bait. Bass bait, walleye bait, game fish bait, are the shiners. Somewhere up in the desolate reaches of South Dakota, there is a bait dealer and there is a population of shiners. The two of them get together periodically, then the bait dealer drives a beat-up old stake-bed truck out over the plains south hauling water tanks and air hoses and shiners, dishing out shiners to other bait dealers along the way. Out there alongside the concrete block building about nine-thirty in the morning, orioles calling and cottonwood cotton snowing down all over the highway in the morning sun, there parks the beat-up old stake-bed truck and the water tanks. Up on the bed is a kid with a net, and out walking around the truck is a woman with her hair tied up, and standing down on the gravel is a man, no, a couple of men, exchanging money and receipts while the truck drips water out on the sand too dry to absorb it. Another load of shiners has just been bought and sold.

"Got any shiners?"

"Sure we got shiners. Now we got shiners, now they's been brought in by the stake-bed truck from South Dakota. How many?"

"Couple dozen."

"That'll be $———."

"They's gone up since last time."

"You can pay the lady inside."

"Where's the nearest place for ice?"

"Over there in the machine. Pay the lady inside for that, too."

"Where's they hittin'?"

"Oh, kinda all over."

"Thanks!" Somewhere down the road there in the front floorboard of a station wagon or a pickup are the two dozen shiners, in their special bucket, off to be bait.

I have some advice and it is this: next time you go fishing with shiners as bait, don't use them all, don't throw the extras out into some watershed where they never belonged or lived, but rather take them home to the kids. Buy an aquarium (doesn't have to be large) and some regular high-protein fish food, and make sure there's some oxygen bubbling through, but make sure you have several shiners, not just one. Now look carefully. Is there more than one kind? How do you know? Is there one with a mouth that is terminal? Is there one with a mouth that is sub-terminal? Are their dorsal fins all the same shape? Do any of them have markings? Do any of them look back at you? Do any of them hang together in the remnants of a school that will never again see the river whose smell they know so well? Do any of them learn within a day that you will feed them? Do any of them appear to have brains?

There are certainly some teachers I know who go fishing with shiners. They put those shiners in an aquarium somewhere then cast that aquarium off in the corner of some room to await the unwary. It sits there, often for days on end, but the fisherman checks once and a while to see if the bait's alive. Then with the patience of a trot-line runner, the fisherman returns to his other tasks, keeping that corner of his eye on the bait. Idle people will always saunter over to an aquarium. He knows that, so waits. Maybe more days pass. But then along comes an unwary student and the teacher sizes that student up in a hurry, deciding whether the student's mind is big enough, a keeper, and if it's a keeper, the conversation goes something like this:

"What's in the aquarium?"

"Notropis lutrensis."

"Do they have a common name?"

"No." (Shiners! Hey teacher! Ain't you gonna tell 'im those is *shiners?)*

"Tropical?"

"They came out of the Platte River."

"These? Why are you keeping minnows out of the Platte?"

"Research animals, those are part of our research into the lives of minnows all along that river," says the teacher with a perfectly straight face, hiding all the questions that teacher would some day like to ask about things that live in a river. So they talk late into the afternoon about the lives of minnows in the Platte, the effects of shifting waters upon those lives, and of the physical alterations of that river that might be responsible for the different minnows' population distributions. And when the student walks out, the teacher knows he's had a nibble.

The teacher also knows the student may be back, especially if that student is what we sometimes call a "keeper," for more and more nibbles, until with a final hard strike, or perhaps with the gentle mouthing of a catfish, the hook is set.

"I've decided not to go to medical school after all."

"Grades?"

"Grades are fine; I just don't think I want to be a doctor."

"Taken the MCAT tests?"

"Nothing wrong with my MCAT scores; I just don't think I want to be a doctor."

"Given any thought to what you want to do?"

"I want to study minnows in the Platte."

Of course day after day in this profession of ours, there are fishermen to be seen, baits to be used, and whoppers to be caught. The game fish are out there by the hundreds and I cast those shiners out into the waters. A teacher's bright ideas, unusual observations, tantalizing thoughts about the world of life, wriggle and struggle through the murky waters of daily life, attracting nibbles, bites, sometimes hard strikes by some game fish not large enough to be keepers. But once and a while there comes along that keeper, and hits that bait with a hard strike, and when that happens I know I've hooked a game fish and I know there will always be a struggle to land that game fish, but I also know that once landed properly, that game fish then becomes the fisherman.

A smart fisherman uses a variety of bait, for there are many species of keepers. There are more kinds of bait available than would ever be found hanging on the racks of some tackle store. And having eaten white suckers, the fisherman knows there are more kinds of satisfaction to be gained from fishing than most fishermen realize. That shiney idea about the control of cell division, that bright thought about the expression of a gene, that observation on the way fungus grows on an agar plate, that picture of the surface of an amoeba, that figure of the structure of hemoglobin, the epidemiology of sickle cell anemia, the embryology of a feather, they all skitter through the gray/brown waters, scattering rays of light out through the particles, capturing the attention of potential keepers. So the teacher wades into those gray/brown waters, casting his shiners all the time into corners. And as he fishes those waters he invariably comes across gill nets from which he tries to free the suckers, and he watches as some of them swim off into their cowboy countries and he is a little sad when others seem never to be able to right themselves, but bend slowly around in circles on their backs until once again they hit the gill net. And as he fishes those waters he invariably comes across chubs, big chubs, little chubs, stuck away in almost every eddy, almost every pool, and he smiles when he sees chubs. He knows they will always find a place for themselves and he need not worry about them.

How it all comes to be only part of a universal process! What such fisherman—wending home in the darkened winter evenings, grimacing against a biting January, shuddering as he stomps snow off boots—was ever able to dry his own lines for his own rest? None, that's the answer—No such fisherman. For when the evening meal is finished, when lingering notes of a daughter's hand at the piano still hang in the air, when the world winds down behind locked doors, that fisherman once again takes gear in hand and casts a line back into his own waters. The choice of books, is that not also maybe the choice of lures with which to troll one's own thoughts? The notes in your own handwriting, made along the margins of some paperback, the page number of a library book, scribbled on your bookmark, are those not strikes from within? The concept that stays through the night and ripples the waters of your pool over morning coffee, is that not the keeper? Yes, yes, yes, those are the lures, the strikes and the keepers, the game fish, trash fish, chubs, and shiners of your mind, in the streams of that inner cowboy country, testing, just testing your own ability, willingness, even courage, maybe especially your own courage to go exploring in that wilderness within.

5

Orioles and Banjos

THERE was a movie many years back, a movie called *Friendly Persuasion.* In one scene Gary Cooper buys a piano, or organ, I don't remember which, but either one represented a startling act of rebellion on the part of Gary Cooper. The movie took place during the American Civil War, and the family of Gary Cooper was Quaker. The seriousness of his musical frivolity was emphasized by some preceding footage in which a huckster tries to sell him a banjo instead. I remember the huckster's line, something about a banjo "stirring the soul." The wild call of the banjo enticed the Quaker, set him up for the piano, or organ, either one representing a major breach of the limits of a mid-nineteenth-century Quaker home. Gary Cooper got in an awful lot of trouble over the piano. There is not much else about the movie I remember, except, of course, the huckster's line. So well do I remember the line that now, as a full-grown adult with an ability to paint myself with a thin veneer of civilization for ladies club gatherings, I own a five-string banjo. I don't play it very well, but then a person doesn't *have* to play a five-string banjo very well. I sit on our piano bench and tune the thing; White Fang appears. I fumble through *Cripple Creek;* wind blows through the curtain. No matter what the tune, the five-string banjo makes only the call of the wild.

Men buy toys; that is a well-known if not widely admitted fact. Men are the buyers of boats, cameras, stereos, fishing gear, and the like. I consider myself an exceedingly responsible citizen, law abiding, loving of family, not rich by any means, but I look around and my environment is filled with toys. The five-string banjo is one. The tape player in my car is another. There was a time when that tape player symbolized attitudes I felt were significant to the ways of life on earth. Its original purchase was made in the face of other needs and it had to be fed right from the start. But it's a different matter now, this business of music and automobiles, for everywhere I go there are cars with stereo equipment. The variety is amazing. There are speakers in doors, in door frames, large speakers lying admist tangles of wires beneath a back window, small speakers in raised black boxes, there are speakers that turn vans into roving adventures in sound. The tape player is always mentioned in the want ads when someone tries to sell a car. "1967 Camaro. Runs good. Needs some work. 8 track. 489–7——." There is this feeling that some kid has been forced to give away his puppy and is looking for a home with the right kind of owner. There is something communicated through this ad in which "8 track" takes on the same significance as "runs good" and "needs some work." I belong to this crowd, so I know in an instant what is communicated. It is the call of the wild.

I travel periodically. A recent trip was to Minneapolis, one of the most civilized cities on earth. Yet by the end of the trip I was a bundle of raw nerves. We'd driven a faltering station wagon without a tape player. Even if that vehicle had had the stereo, five people simply can't tolerate the same kind of wilderness experience that a single person, alone for hours on the highway, can. For days in a big city hotel, wandering the curves of Nicollet Mall, taking pictures of my son atop the Walker Art Center, the IDS building miles away held gently in the fingertips of abstract sculpture on the roof behind him, I drifted through the high culture that was Minneapolis. Everywhere there were well-dressed people. Classical music was in the air. Television commentators seemed too intelligent for their jobs. Everything was good. Almost. There was no call of the wild.

I went into withdrawal. My ears ached for Waylon Jennings, pounding, earthy, free; for sawing away high hoarse Bill Monroe; for precision Earl Scruggs; for the rangy Nitty Gritty Dirt Band making mockery of limits. After a few days of this, I got desperate and went to the Minneapolis Zoo. Salvation swam in the beluga pit. *Real whales.* I thought immediately of an unholy mixture of country music, whales, orioles, banjos, and a spacecraft billions of miles away carrying the sounds of our world and diagrams of our kind. It is time to admit that I had thought many times before about sensory mixtures such as these. But it was the sight of whales swimming the high civilization of Minneapolis, combined with the incident with the oriole, that finally told me the significance of these thoughts. Let me gather up some loose ideas about raw music and whales, for then the actions of this single oriole will be revealed for what they were: a communication between the wilderness and a human being. This communication, played out in the medium of the five-string banjo, would never have happened, had I not been preparing for a month and a half in Keith County.

Every year for the past several years my family and I have moved lock, stock, and barrel to the West, to a town that boasts a "Boot Hill," baseball as only it can be played out there, and a natural wilderness in which I become totally immersed. This year I vowed to prepare properly, to be organized. One doesn't head out onto the interstate west without a proper supply of music to feed the stereo, properly arranged on cassettes that can be reached easily at highway speeds, music which sets the tone of the weeks ahead. Most of my tapes were hodgepodges. Waylon here, Willie there, Tompall in between, Flatt and Scruggs mixed with Ronstadt. I started in the morning re-doing all this mess. A grown man playing with his toys, you do understand that's all it was, right? The wind blew gentle spring rustling ash leaves, the sun was bright, and things were slow outside. Then came the incident with the oriole. I need to give you some more background. Through the following paragraphs there is a thread of a tale about how a human uses and views the music at his fingertips. You probably need to have that background before my communication with a single oriole will be placed in its proper perspective.

One of my tapes, played frequently in the car on a cold morning in the dark, has recorded the songs of whales. The songs came from a *National Geographic* record, inserted as a page in the magazine months ago. These whale songs are among the earth sounds recorded and carried by a Voyager spacecraft, a craft that is destined to wander the universe. It is entirely possible that these songs will survive their singers. The first thing I did with the *National Geographic* record page was to put it on tape. The next thing I did was to put Waylon Jennings on the rest of that tape. The next thing I did was to put the whole thing in that slot in the dashboard. Humpback whales groaned in my car, long eerie resonant groans; I could feel the waters of Bermuda. A drawn out singing scream, then gratings, then more groans, all echoed off the water surface above, off the rocks and corals below. Shapes moved through the traffic; a long barnacled flipper passed across my windshield, then they were gone. The songs were speeded up fifteen times and became bird calls. There was a mockingbird, insistent, polyglottic, confident, in the back seat. It was at that point that I decided there *is* a universal language in which all living things can speak. But that language is music and the grammar is inflection, speed, and context. I have little doubt that whales could converse easily with mockingbirds, orioles, or the country musicians at my fingertips, for they all play the sound of the planet.

There is an old Johnny Horton tune, "The Battle of New Orleans." But the modern Nitty Gritty Dirt Band version of that same music begins with the songs of humpback whales. Groans and drawn out wavering set the mood for what happened down under those dark trees in southern Louisiana way back in the early days of our nation. Once that observation was made, then the rest became obvious and the oriole, well, the oriole probably had no choice but to communicate with me through this medium of country music. Once the observation about the Dirt Band and whales had been made, then I was sensitized to the para-music communications that wind through popular culture. The miles rolled away beneath my floorboards and the tapes rolled away at my fingertips, gritty winter city miles and grittier musicians, and I heard the sounds of nature: the calls of whales, the calls of coyotes, the squeaky buzz of a wren, all at different speeds. Blizzards blew pasting sleet on my windshield but inside there pounded the wings of cranes, beat by some group that never knew a crane, but only knew the cranes' wildness within themselves. The winds came from the north and with vengeance, crept in among the cracks and played weather stripping. But as I pushed harder on the door there came the knowledge that out there on my driveway, inside a rusting box of steel, there was a place I could put my finger and push. The cassette would go in, the wheels down inside some Japanese technology would turn, and someone would bend a harmonica to play the song of the wind through that weather stripping. You must understand there is a synthesis going on here, in which the sounds of rebellion as manifest in the infinite sea of American folk music become one with the sounds of wilderness as sung by the planet.

It has been a long time coming to this, this final admission of retrogression. But all my close friends know it's true so it can be stated publicly: I am driven somehow to those places in all of nature that are wild. This is no contest with the forces of nature, no struggle for survival, no "use your wits or die" circumstance. No, this is instead a celebration of the free thought. This celebration may only be the early forays of this wild planet into my consciousness, testings of the communication system. Do you hear the sound of planetary winds in the harmonica behind Willie Nelson? Yes. Can you hear the whipping pounding wings of cranes in the guitar behind Waylon Jennings? Yes. And can you listen to that guitar and then be instantly in that muddy field behind some barn where you once stood and watched cranes as far as your eye could see take to those pounding wings? Yes. So the inside of my car is one of the wildest places on earth. It is a science fiction chamber in which a push of the finger can take me to wilderness, a wilderness of the mind.

Finally, I should tell you the instance of the oriole, for that one instance is the basis for my final admission of the nature of this kind of music to which I must listen. I was preparing for Keith County, it was late spring, the windows were open. I was taping my newest music, Bill Monroe, in preparation for weeks in the hills. A supply of proper music is required equipment for weeks in cowboy country. Bill Monroe is proper. The tape turned, the turntable arm dropped, and into the morning suburban air wafted the most basic of American sounds: someone on a banjo calling Bill Monroe. And in response to the communication, into the morning suburban home wafted the most basic of American sounds: northern oriole, throaty, insistent, confident, simple, *close.* I tried it again and it worked again. Close, and coming closer! Called by a banjo, called by a fiddle, it called back with the para-music of wilderness and free thoughts. Calling J.J., calling J.J., are you coming J.J., into that state wherein lives the oriole? Yes I am coming! Cowboy country! And furthermore, when there's no chance for real cowboy country, then to a cowboy country of the mind. And when I'm gone there, I cain't be reached by no telephone, and if you send a memo, send it rural route, and if it's not important, if it interferes with my thinking, then send it to my boss instead, or maybe to *his* boss, 'cause I sure as hell ain't gonna read it! And if we *really* need to reach you, off in the Keith County of your mind, off with your own thoughts? Find an oriole, or a banjo, or a harmonica, fiddle, whale, something that makes the sound of the planet, and play the music of Earth.

6

Love and Love/Hate

LOOKING back over many of the last twenty years, it seems my life has been tied up with vehicles more than that of the average citizen's. The auto and its relatives and subspecies have never been anything but the barest of necessities for me. I have no surrogate love affair with a *brand new* car. In fact, probably because of economics, I have come to be more comfortable in a clunker. But I must travel as long as I can the highways of these plains; there are things I must see in the grass, in the mud, and on the sand. So my life has been tied up with automobiles. So much so, it seems, that one of my more common positions is bent over peering under the hood, smoke or steam billowing as the highway traffic roars a few feet away. I love to be out on the highway headed for the country; I hate to be bent over under the hood. Love/Hate.

There is a pond out in Keith County; it goes by the unofficial name of "Mudhole #1." It is one of the most unique ponds in all the world. I have studied Mudhole #1 for six years now, and I can say this about it: it is about the size of your living room, it is appropriately named, and I have never failed to be floored at what I find there. On one particularly interesting trip, I found there a pair of mated western grebes, one of the wildest of species. They could not be disturbed, would not fly, or dive, would not leave, would do nothing but move slowly and gracefully within feet of me, posing, even staying on Mudhole #1 while I went into town for more film. Love.

That expression of grebe love was one of the most powerful experiences I have ever had with wild things, and heading home, my companions and I could talk of nothing else. Then the state vehicle, a van, exploded. That particular van was my favorite state vehicle; you'll know why later. But what followed the explosion was a set of experiences that, in my mind, tie one vehicle, the entire automobile industry and automobile service industry, permanently to that most powerful of wilderness experiences with those western grebes. Love and Love/Hate.

There are primitives in the world who would be able to analyze this Love and Love/Hate connection easily, for they purposefully, culturally, socially, practice similar juxtapositions of seemingly unrelated events. Some anthropologists have decided they can also explain the purposeful juxtaposition of unrelated events. Well, I've read what the anthropologists have to say on this matter, and I've lived the primitive experience out on the highways of our land, and I can tell you that the anthropologists are correct. But no primitive has a Mudhole #1, and no primitive has a state-owned van lying in pieces out on the interstate, which of course is a shame. The combination would relieve their needs for a whole spectrum of puberty and other coming-of-age rites. I can see it all now: off in some jungle stands a near-naked elder, his arm around a young boy on one side, a young girl on the other, and in front of them passes a natural phenomenon so unique, so powerful, that it can only be an omen. The natives are exhausted; they've stalked the trails for days, maybe weeks, sensing that this event lay ahead. But the elder is so wise in jungle ways; so guides them unerringly to a place he knew *could* produce the natural omen. Suddenly that omen is there and will not go away. The elder explains his culture's ways, means, mores, and responsibilities, as the youths stand mesmerized. They turn away from the omen. Waiting behind them in the jungle is a

1975 van with "Property of the University of Nebraska" stenciled in white on the door. The van self-destructs. The elder, now with scolding, incisive tones, directs the children to reassemble it. The children do the best they can, with the omen and its associated lecture still in their minds. It is an experience so all-powerful that it sears new openings in their brains. They never forget the natural omen. They never forget their mores and responsibilities. They never again feel the same about the van or all it symbolizes.

Maybe I should expand on Mudhole #1.

It's down the road where the swallows nest, it's beneath the sand mountain, and it's in among the oriole trees. The oriole trees are chapel stillness, and clear of underbrush, with the throaty fruit of an oriole's brief song echoing. Gravel dust hangs there on insecty mornings. The water looks poisonous; it may be. There are cattails by the road, on the north and south corners of Mudhole #1. The snail population of the southeast cattails is not of the same makeup as the snail population of the northeast cattails. The two cattail stands are twenty feet apart. The water below the cattails is crystal, but with a greenish tint. Clouds of aquatic plants billow beneath the surface. There are a few places you can see green mud beneath plants. There are a hundred thousand snails in Mudhole #1. It's the kind of place you would never have waded into as a kid; your parents would have put the fear of God into you. It's the second place I waded into years ago, as an adult, experiencing for the first time the heady euphoria of cowboy country. The mud beneath the water goes down a long, long way. Nowdays I try to avoid the poison ivy on my way into the mud.

How many times I've done this. That's always the thought when I put the state key into the state orifice and activate the state battery that turns over the state engine of that state van with the state tires and the state sign on the state door. Didn't Tom Wolfe have some way of describing this feeling a civilized person has when he participates in a barbarian act? Some urge to return to the primeval scum surfacing in some culturati's *demeanor, act, fad. Is that not a bubbling described by no other than a Tom Wolfe? The state engine grinds, and in that sound is the gurgling of primeval ooze.*

The van side doors slam behind me; the inside handle is pounded downward into place. A smile creeps across my face, the smile of the barbarian. It's all right to be a biologist after all! That's my final thought. That van always clunks horribly, with a sound of metal tearing metal shuddering up the differential through the transmission and into the floorboards, when you put it into reverse. I love it. An element of chance suddenly sparks in the air; there is a brief smell of sulfur, something burning.

We'd seined Wellfleet. "Seining Wellfleet" has become an expression for that entry into the hyperspace of field scientists. Ominous heavy mechanical robot breathing echos through that hyperspace, and heartbeats, amplified, fed through speakers, pound in your inner ears in that hyperspace. There is nothing to seine at Wellfleet. Nor was there anything to seine at any of the other dozen or so places we'd been. Late spring, that's when it all was, on a rare day in the west. The day had started at 2:00 A.M. We'd waited for daylight beside a hog lot by a flooded, vengeful river. We'd ended with Wellfleet. There was a little building there, its screens falling, and a darkened window in the door.

"Wellfleet has a bar," said someone. The place was closed, fortunately. Had it been open, we might be there yet seining after nothing. Peter Mathiesson—he's the one who wrote most powerfully about seining after nothing. His book is *Far Tortuga,* the most ethereal of tales of man's quest for something no longer there. No need to go to the Caribbean, I thought, raising the dust of Wellfleet, you can seine for nothing right here in Nebraska. We built a fire in the cabin in cowboy country. We talked late of sophisticated things. We ate cookies and drank coffee and threw our cigars into the fire. Sightseeing tomorrow. We'll start with Mudhole #1. With a little luck we'll see bald eagles on the way. Tomorrow dawns. I wish every human could wake up one morning in May at 5:30 A.M. in Keith County. Eagles gone. Mudhole #1, we're on our way!

I was not even concerned with the timing chain, I was concerned with the front end. It never steered right. I drove it a couple of thousand miles that summer, then drove it a couple of thousand more back and forth to places like Wellfleet, then to Oklahoma a couple of times, and it never did steer like a van is supposed to. I had some concerns over the front end. It went back into the shop for the fourth or fifth time. I wouldn't drive it another summer. I simply refused, called them on the carpet, spouted off about liabilities, accidents, what would happen if the thing fell apart out on the highway. In the end I was the one who had the accident. It all had something to do with worms in Oklahoma. In the end I was the one driving when it fell apart out on the highway. So it went back into the shop. Then it came out a last time before summer. We were going through this elaborate ritual of deciding who drives which van and I took it. My assistant drove the new one, the brown one with a radio, nice seats, girls. By then the front end was fixed.

"They put a stabilizer bar of some kind on it. Should drive better," said Mel.

"It handles a lot better, but it doesn't seem to have quite the power it did before. Seems a little hard to start, a little weak or something."

"That was a re-call item, the timing chain. Some of the others complained about it stalling in traffic. They tuned it up. Didn't do much good. Last time I had it in they admitted it was a re-call item, the timing chain. Or something about the timing mechanism."

"How many times did you have it in before they told you it was a re-call item?"

"Several." He'd been a Navy Warrant Officer. None of this was strange to him. He knew a lot about government property.

"Thanks, Mel."

"You're welcome, Doc. Hope you don't have no more trouble with it."

"Wouldn't be a trip without trouble with this van, now, would it?"

"Guess not." He chuckled.

I stuck the state credit card in my pocket. It was raining outside, damp cold spring raining, light rain. The van had been closed, locked, for a couple of days. Someone had years before lined the inside with Masonite pegboard. When it's been locked up for a couple of days, and it's been raining, the Masonite smell permeates everything inside. Your field clothes smell like that for weeks. It's a sickening smell. I've come to associate that smell with western grebes.

* * *

There are ads on television these days, ads showing people taking pictures with motor driven single lens reflex cameras. "Just focus and . . .," go those ads, and that special sound of the motor-driven SLR clacks and whirs in the background. I've been the subject of some photographic sessions myself, in the last couple of years, so that motor-driven sound, clack-whirrrr, clack-whirrr, is familiar from real life, too. The amounts of film you can consume in such a session will stagger your mind. No frugality here in the motor-driven SLR business, no indeed, the celluloid goes spinning! I don't have a motor-driven SLR, but the temptation is there, if only for the sound. Maybe that special sound would make me the photographer I want to be. But in all my associations with the wild, in good places and bad, wet and dry, usual and unusual, I can think of no place more suited to a motor-driven SLR than beside Mudhole #1 on that crisp morning in May. I was out of film.

My camera lay on the engine cowl as we stared from the van, scarcely breathing, three grown adult American males, oohing and ahing over a pair of grebes twenty feet away, and I was out of film. Cars passed, raising dust, drivers glancing over as they tried to see through the weeds in the direction we were looking. Dust settled; they were still there. Total grace. Total wilderness. Totally in love. The air was charged with grebe. Town was ten miles away. The grebes—their bodies were on Mudhole #1 but their minds were on another part of the planet. I ground the van into life. They'd be there when we got back from town. I've been in that blissful state before, that same sealed-off condition as those grebes. In that condition one never thinks very far ahead, nor analyzes. I drove into town like a bat out of hell anyway.

"Darlin', we've been down some roads together; the Lord knows that they ain't all been too smooth . . ." goes a lyric from Tompall Glaser. No one could have said it better. The images run together now, there have been so many of them, in so many places, with so many people over the years, and all of them good. They all occur outside of town. It's in Kansas, cold, rain,

sleet, on the way to Oklahoma for worms, and the defroster doesn't work. Or at least I can't figure out how to make it work. We wipe the windshield for three hundred miles. I stop for coffee; they kick us out of a roadside cafe. The van has "Property of the University of Nebraska" stenciled on the side. The University of Nebraska has just defeated the University of Oklahoma in a football game. That'll get you kicked out of a roadside cafe in Oklahoma. Stopped for gas in Norman. Little kids come up and start kicking the hubcaps. I raise the hood. There's a wire under there that doesn't look like it should. I push on the wire. The defroster works.

Then there are the windows; they always get broken. It's these roads, some say, these washboard roads are what break the windows. Then there are the side doors, never quite closing right, always slamming. Then there's the lost key, lost somewhere in the North Platte River. State never issues duplicate keys; you're not supposed to lose a key. Are you supposed to wade in the North Platte River? Ever see a van towed at highway speeds? Interesting. There are also guys who can get into any vehicle, then make a key for the lock they've burgled. Even in Keith County there live guys like that. Then there's the accident. She was beautiful, delicate, petite, and I felt sorry for her. Her husband must have been very impressive; she was terrified. She was driving a Chevy sport pickup with about ten miles on the odometer. She'd gotten it yesterday. You could tell it was her husband's favorite toy. You'd think she would be her husband's favorite toy, wouldn't you? We searched, we really did search, as very diligently as a vanful of scientists can search, for a mark on her bumper. There was no mark; the collapsible bumper worked perfectly. There was all this serious discussion and the clacking of cameras in the background. I never saw any of those pictures, evidence of my only wreck in a state vehicle. I think they were all taking pictures of her. All I remember is her Okie accent. It lingers in that van, lilac lacing wet Masonite.

Then there's the van parked on my driveway at home. My son touches it, reverently. A misty look comes over his face. "I love these vans," he says. He's nine years old; for five years these state-blue vans have symbolized Cowboy Country, U.S.A. Then there are the fights over whether some filling station would take a state credit card. Boy, did I have one coming-up over a state credit card! But always, it seems, always it's me out there on *the front lines, and always, it seems, there are some laughing yahoos sprawled all over the back seats. Kick the bunch out, send 'em out to pose for a Coors commercial! It was May, and dark, late at night. The van sat in our driveway. There is a sense of anticipation, excitement, about a van on your driveway packed with sleeping bags, seines, and all that Keith County gear. I am so cunning. I took the tool box out of my personal car and put it in the van. Ever get the feeling that someone somewhere, some unseen force, is telling you something? I must have had such a feeling. I put the tool box right beside me on the floor. Reach it in an instant that way, right? Right.*

Wild, that's what a western grebe is, pure wild. You can never get close to them in a canoe, out on the big lake, they always drift off just when you think you're making some progress. Sometimes they'll go under, sometimes just sinking down deeper and deeper until they're gone. Sometimes it's a dive. Then you paddle like hell hoping you can get closer before they come up. It never works; they're always further away and in a new direction. I don't think I've ever seen one fly. On a still day they leave long vees of ripple. There are plenty of times I stop what I'm doing and just watch western grebes from a long ways away. A flood of uniqueness comes over a person watching western grebes: the whole world stands still—the wars, the energy crisis, local politics, athletics, it all goes away and it's just you and the grebes way out on the lake, fog hanging down between the bluffs early in the morning, and the grebes way out on the lake making their long vees in glass water.

Found a dead one once, up on the lake they call Big Mac. It was summer so I can't believe it would have been shot. The death of that bird has always been one of my mysteries of cowboy country. I don't visualize western grebes dying, especially in this part of the world. Sometimes, though, I visualize the rest of the world of wars, malnutrition, invasions, and totalitarian states dying, and the western grebes surviving, heading down through the eons of time making long vees in still water. Maybe that's the same vision I have of all that can be called "pure wild."

At our time of the year, they're usually paired, or at least go around in pairs. The most noticeable thing about a pair of western grebes is their heads; well, the heads and the necks. Of course "paired" is too weak a word. They do seem paired, in a sense, because they're together, and they *act* paired, that is, paired in the reproductive/biological sense. But "paired" is too weak a word. "Bonded" is better. But still better would be some word that could connote a parapsychic, telepathic, outeruniversic, molecule/radiation field, a merging into composite orbitals, a charged space in which the electromagnetic waves stirred up by the turning of one grebe's head then automatically turn the other grebe's head. Back from town I loaded my camera, still trying to think of such a word. We all got out and walked to the shores of Mudhole #1, the western grebes gliding at the points of their vees, ten, fifteen feet away, their charged space, their parapsychic, telepathic, extrasensoric electromagnetic waves merging our own into a grand set of composite orbitals. Even Mudhole #1 emitted waves. I could smell them. Cameras clacked and clacked. We separated, one of us moving to the other side of Mudhole #1. The grebes came to the middle, never more than a foot apart, a *pas de deux,* an omen, a wilderness experience not granted everyone, a kind of wilderness experience from which some Indian might have derived a name. I wanted that Indian name: John Seen-the-Grebes. It went on for an hour, two hours, and the film was gone. We were back in the van in silence down the road before someone came up with the word I wanted, the word that connoted all that merging into composite orbitals.

"I think they were in love," someone said.

It all seemed so logical, this decision not to shave for several days. But in the end I had to admit that maybe, just maybe, it was that logical decision that caused all the trouble. Maybe I was only kidding myself about the logical bit. No decision made out of vanity has ever been really *logical, now, has it? And the decision not to shave for several days was pure vanity. It looked so bad after the first day that I had to leave town. So we threw our seines in the van and headed to Franklin, Alma, all those little towns, and even circled "Wellfleet" on the map. Two* A.M.*, that's when it started, officially. We were on the Republican River south of Franklin before daybreak. Had to wait for the sun; kept hearing all this clanging and banging off in the dark. We'd come to seine a hog lot in the dark, and the hogs were banging their feeders or something; very regular. It was all slop in, out of, and around the hog lot. The river behind was all slop, on the downside of a flood. It went downhill from there. Somewhere west of Franklin I asked permission to seine a creek. After a long time some guy came charging like a wild bull to the door, buttoning his pants. From off in a back room came female laughter. We got permission. Know what I remember most about that creek? The wrens singing off in some dead trees, that's what I remember most about seining Fox Creek. Was it right before or right after Fox Creek that we found the red, white, and blue mesh cap on the road? Right after, I think. Let's see, where was the dog pack? Some place we decided not to seine. Cattle trucks? They were south of Wellfleet, headed north, a hundred miles an hour. I would swear on a portable aerator they were going a hundred miles an hour. I was going nearly eighty and they passed me with a whoosh. The visions merge. The vision that doesn't merge is my face with three day's growth of beard: the one logical decision, made of vanity, that came back to haunt me. Funny, all those places we went, all those sights we saw, all those terrible biting flies under the railroad trestle east of Wellfleet, all that disgust at the Wellfleet bar being closed, funny how most of it merges into nongrebe memories. Grebe and nongrebe, those are the classes of memories from that trip. Well, I take that back. There is one more class of memory. It has to do with state vehicles.*

It's a sound you don't forget—the sound of technology in anguish. Threw a rod, that was my conclusion, a piston rod. I had visions of the rod coming up through the floorboards. There was a smell. I could almost feel the fire billowing up between my legs out on the interstate. Then we were on the grass, engine cowl thrown over in the ditch, cameras retrieved in case of fire, cop wheelin' it across the median. The Paxton turnoff was two miles away. Threw the cowl in the back seat and headed out. You know how vans are built, don't you, with that engine cowl coming back almost up into your lap? It was raw engine all that day. The last memory I have is of that raw engine clattering and the naked interstate through that gaping hole at my feet. The temperature started up as we hit the exit ramp. Lot of mystery went out of that adventure right then. I've been in lots of cars when the temperature starts up. We coasted down the long hill to a service station. Alternator bearings were frozen, fan belt broken. Bought a couple of fan belts and shot some oil up into the bearings. Interesting thing about that kind of van is that you have to take off the power steering belt to put on the fan belt. It was grease here, grease there, grease, grease, everywhere. Add all that grease to four day's growth of beard. Paxton was closed down tight; noon hour. Headed back out on the interstate. First fan belt lasted about ten miles. It was another twenty to real civilization. More grease.

"Of all the stuff that could have been on that little ole pond," someone said, "it was western grebes."

There are, I suppose, some things a person never forgets.

They must fly. There is no statement in the literature that they walk on their migrations, so they must fly. But I've never seen one fly. Secure on a lake, there is no way to make them fly. So many others, it seems, have no choice; something in their behavioral patterns tells them to fly, but not the western grebe. The western grebe paddles. The western grebe dives while others fly; the western grebe sinks slowly to any desired depth, leaving only that periscope of a head turning flatly to the right, the left, while others fly off, pattering across the open waters east of the cattails. There could not be these long subterranean murky water-filled tunnels, could there, through which western grebes swim upon their migrations? No, there are no such tunnels, so they must fly. We tried to make the pair on Mudhole #1 fly. Get serious. It didn't happen. In defiance of all normality, they stayed, never a foot apart, moving only a few feet away, in defiance of all human antics designed to make them fly. They had no time to fly; they were in love. In retrospect, we could easily have pushed their love to the ultimate test: wade into Mudhole #1 from three directions at the same time, actually try to wade up and catch them with our hands. The morning was too bright, the night before too filled with conversation, the dry jeans too comfortable after Wellfleet and the Republican drainage, the air too warm and charged with spring after the most brutal of winters. Not even the most hardened field man wades into Mudhole #1 on such a morning. Accept the experience as an omen, that's all there is to do with western grebes on Mudhole #1. I wrote some scribbled notes, sitting in the van with road dust lingering, but it was wasted activity. My field notes are usually rudimentary at best, designed only to remind me of the sense of a feeling had at the time, of an earth-shattering idea worthy of pages of manuscript. But on that day the notes were wasted, for down the road by Paxton and beyond lay a set of experiences with modern technology's service industry that would forever seal those grebes in my mind as the ultimate wilderness message, the ultimate statement of primeval love, the ultimate communication, the ultimate gift of nature.

Jesus is always pictured with a beard, right? So are Darwin, Matisse, Leonardo da Vinci, Lenin, Kenyatta, U.S. Grant, Joseph Conrad, Anton Chekov, and Rasputin all pictured with beards. So what's wrong with J. Janovy, Jr., with four days' growth of beard? You see this decision not to shave was the most logical of my recent ones. It was time to grow the annual mustache, which in the summer serves admirably as a sun-shield. My lower lip is evidently too thrust out at the world: it catches all the rays. So I grow this mustache every summer to keep off the sun. It works like a charm. But there is nothing more stupid looking than a forty-plus-year-old man with a few days' mustache. So all you have to do is grow the whole beard then shave it all off except the mustache, right? A forty-plus-year-old man with a few days' of all-over beard, now, that's a fairly common sight. You see it all the time down on a couple of corners of this town. I guess you don't see it too often out in cowboy country driving a broken state vehicle, covered with grease, shirtless, demanding alternators be charged to state credit cards. I guess that kind of sight elicits allergic, hyperimmune, psychological anaphylactic shock in certain kinds of businessmen out in cowboy country. The kicker came many hours later, middle of the night, in my own house.

"I'd have kicked you out, too," she said, and she's been married to me for almost twenty years! Anyway, we talked about it in the lab for days afterward.

"I made him mad right from the start."

"You made an 'F' in public relations."

I'd known that: a state credit card, the total frustration of that splintered van, some guy half my size, my good sense not to start a real fistfight, all that city mission beard, all that grease, three hundred miles from home and no hopes of ever getting there, a tantrum, some screaming about cutting off all the business the state does with that outfit, some more screaming about how they were doing the state a favor by working on their vehicles (don't ever in a million years believe that), finally all that stuff about the Indian Agency. How did we ever get off on the Indian Agency? Easy.

"But I'm a professor at the University of Nebraska!"

"I don't give a damn who you are, you ain't gettin' nothin' out of this place with that credit card. Guy from the Indian Agency came in a month ago, had a card just like that, bought a bunch of stuff, we never got our money. Get the hell out."

"But I'm a full professor at the University; this is a University credit card, not the Indian Agency!"

"Get the hell out."

You want to know the final bit? Six months later I get a call from Auto Pool. They wanted to know why I'd bought two fan belts at Paxton and charged them both on the state credit card. But Paxton was way back down the road from where I sat in four days' growth of beard and only the memory of grebes to trade for an alternator.

"Get the hell out!"

We got out. Somewhere out on the highway near the town of North Platte, Nebraska, there is a Standard service station. If you're ever in that area, stop by that Standard station, will you? The man there is a rare soul. He got in his pickup and went into town and found us an alternator. Then he charged it to the state credit card. Told him I was a professor at the university and some day I was going to be this famous writer and then I'd *tell everybody to stop by his station, give him the business.*

"Here's your receipt," he said.

It was late, very late, in the darkness of late spring nights out on the plains, and the wandering cars tracking their ways home from prairie taverns plied the streets of my town alone. I slammed the van door. Any valuables such as they might be would still be there in the morning, thrown under a seat. Nobody breaks into a van with "University of Nebraska" on it. But the real valuables, the western grebes, would also still be there in the morning, tucked away beneath the seats of my memories, and they would still be there the morning after that, and after that, and every morning after that forever and forever, as would be the auto agency in North Platte. The front door was open. I stormed in with all the words you've just read about grebes and technology. She listened with sleepy interest, but then summed it up with, "I'd have kicked you out, too, Professor."

"But if you'd seen those grebes, you'd have stayed to watch."

"Yes," she said, "I'd have done that too."

7

Tigers and Toads

THE Rocky Mountain toad may be one of the world's truly ugly creatures, also slow and ungainly, showing no outward sign of intelligence, and reproducing in obscene numbers on the sandy shores of Arthur Bay. The tiger beetle may be one of the world's truly beautiful creatures, also exceedingly fast and graceful, running as well as flying, exhibiting behavior suggestive of style and class, stalking the sandy shores of Arthur Bay. There could be no two creatures more illustrative of the extremes to be found within the animal kingdom. Yet, they spend their days in the tightest of life and death relationships: the toads eat the tigers, and in gluttonous numbers, right there on the sandy shores of Arthur Bay.

Surely there is some lesson I am supposed to learn from this! was my first reaction to the above observations. The occasion was routine enough: a tapeworm had been discovered in toads of Arthur Bay. Tapeworms of land animals often must exist for a time in lower animals, insects, for example. The insect is eaten by the land vertebrate and the worm is freed to continue its development into an adult in the vertebrate's intestine. I don't remember the origin of that move to examine the stomach contents of the Rocky Mountain toad, but that move was obviously motivated by a curiosity about toad diet. In retrospect, I do strongly wish I could remember that moment I decided to look into a toad's stomach. That is a moment I would like to repeat again and again at other times, in other arenas, in other laboratories, with problems other than toads.

Some combination of events, of circumstances, of people there at the moment, some spark of interest struck by a previous event, all worked together to generate the question: Wonder what toads eat to get those worms? How I do wish I could put together at will a set of circumstances that would produce similar thoughts about other things! Oh, I've generated plenty of curiosity about those other things; the generation of curiosity has never been a problem. The problem is the generation of a curiosity that always asks a question that in turn spurs one on to a physical act in pursuit of an answer. And I do wish this physical act of observation would, not just once in a while, but *always,* have the lasting power of the sight of a toad's stomach full of tiger beetles.

It was years ago, very near the end of our annual stay in Keith County, when I made the observation about tigers and toads. But at that very moment, the first paragraph of this essay was written in my mind and has stayed there ever since with virtually no change in wording. You see, tiger beetles have a certain reputation among those who know them only slightly, or have tried to catch them. The reputation is one of speed coupled with almost purposeful behavior, a sort of halting run, that makes a tiger beetle simply very hard to catch by normal means. One does not walk up to a tiger beetle and pick it up. Nor does one easily sweep up a tiger beetle with a net skimmed over the white blistering surface of Arthur Bay dunes. No, the way to catch a tiger beetle is to act like a toad. The way to catch speed, grace, beauty, elegance, is to act dumb, slow, ugly, ungainly? No, patient. The Rocky Mountain toad is a packet of patience. But there is something deeper here: Patience is nothing more than the correct approach, and therein lies the lesson of this beauty and this beast. *The* correct approach, regardless of outward appearance, is the one attribute that will always net the prey.

And into how many other circumstances can this rule be inserted? Into many, is the answer. On many barren dunes will the correct approach capture beauty and grace. All one has to do is develop *the* approach. Now there is a challenge! We are not always born with it, latent, awaiting only time, warmth, and water, as are those black polywogs out near Arthur Bay. But we can plan for the correct approach. We can develop the correct approach on purpose, or teach it to others. There is more beauty and grace than that of the tiger beetle awaiting! You do understand that I am resisting the

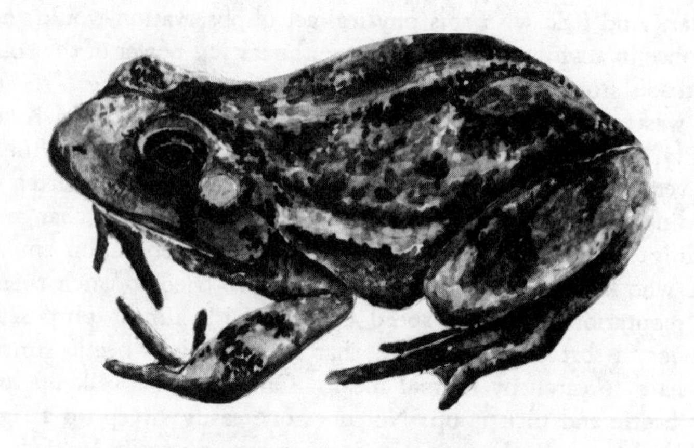

". . . slow and ungainly, showing no outward sign of intelligence, and reproducing in obscene numbers on the sandy shores of Arthur Bay."

drive to preach too early a personal sermon: The lamination of early decision, of sealed-off options, of the euryphobia attending false security, elitism, or lack of confidence, are all the things that prevent us from accomplishing what a Rocky Mountain toad can do. No, I'll save that sermon for later, if at all. Instead, I will extend this tale into the absurdity of toadal aspirations. Toads, regardless of their beastly countenances, have access to beauty and grace, but not to the beauty and grace of the swallow. Can you not envision that cowboy toad, baking out on the flats of Arthur Bay, watching barn swallows above, watching *their* beauty and grace that more than equals those of the tiger beetle, and saying to himself (or to another toad) "One day I catch *that.*"

Let us shift for a few paragraphs to a discussion of *Rana catesbeiana,* the bullfrog, not a toad, but a relative of the toad. Then we'll come back to Mr. *Bufo woodhousei,* dry land cousin of *R. catesbeiana,* the green, waterlogged, euphagous bullfrog. You've tasted bullfrog legs; they're delicious, sort of like rattlesnake, better than chicken, sort of frog-leggy. Around here, the bullfrog is big, capable, on the lists of game animals with *seasons.* You must know that *seasons* set the bullfrog apart in the mind of a human, apart from the other anurans of cowboy country. To me, however, it is his name that sets him apart in a way no season could. *Catesbeiana*—now there is a name of all names. I have no idea who Woodhouse was. I'm sure he was famous in some circles. But I know the name Catesby well.

Mark Catesby was a naturalist, antedating Audubon, and by all standards of modern cultural ethics, was a bum. He came to the United States of America on the bum, married the daughter of a wealthy landowner to support his bummery, moved onto his father-in-law's land, and spent the rest of his days in the woods of what is now North Carolina, collecting plants and animals, drawing pictures of the specimens he'd collected, sending those pictures to the leading biologist of his day, a fanatic named Carl von Linné, and generally pursuing a life of pleasure. This was all done without malice, according to his biographers, but not without forethought. Nobody, to my knowledge, finances a scientific career in that manner today. But then science has grown large, sophisticated, somehow apart from the run of regular society. So I've never heard an aspiring young biologist admit outright he'd like to find a wealthy young landed lady to support him so that he could pursue his studies. But some have come close.

The point of this discussion about the bullfrog, however, is that upon its list of edibles is birds. Now I find myself wondering if there is not some anuran interlanguage, some set of croaks by which *B. woodhousei* could be told that upon the list of his cousin's edibles is birds. Modern biologists tell me of the impossibility of such interlanguage. Toad calls, frog calls, are wondrously specific, say these folks, and indeed recorded anuran calls, and their respective answers when those recordings are played out in the weeds, are stock in trade of those who study "isolating mechanisms." Such a shame that the independent and asocial anurans should live with that separation of language, that dearth of intercommunication, and in so doing should not be able to tell a single *B. woodhousei* who would have a swallow of swallow that the lofty ambition was indeed lofty, but at least worth a good try! Such fortune, that in the life of that most social of animals, *H. sapiens,* whose interlanguages infect all layers of society, the binding communication between individuals of different ilks is the very element that opens the *idea* of opportunity, that passively sanctions the lofty individual aspiration beyond which immediate companions would strive.

I tested my limits one day, out on those white sands of Arthur Bay. I took upon myself the task of getting close to a tiger beetle. The idea that morning was to establish a photographic record of the tiger-toad relationship. That relationship would make a good subject for a charming lecture to some civic group. After all, Beauty and the Beast has stood the test of time; and tigers and toads is a sort of beauty and the beast kind of tale. But the tiger proved more elusive than the toad, the beauty more elusive than the beast, until, of course, I acted like a beast. So here is a short biology lesson for those of you who will never crawl around in the Arthur Bay sand: In order to photograph the tiger you must be patient and slow, always giving the beauty time to adjust to your presence. You must not all at once present too much of a difference from your surroundings. And, it helps immeasurably to get down on the same level as the beauty. With those simple techniques, *the* correct approach, you can come as close to beauty as your beastly desires want. I was too close for my lenses. I lay in the sand of Arthur Bay eye-to-eye with a tiger beetle, and successfully resisted the temptation to eat it, sensing, somehow, that my tongue might not be as sticky as a toad's. I stood then, to full posture, brushing away the sand, and beauty flew away many yards. But my eyes had adjusted and suddenly there were hundreds of tiger beetles everywhere, ones I had not seen before getting eye-to-eye with one.

I'm sure the recreational folks of Arthur Bay saw me as different from themselves that day. And I'm just as sure that my behavior required no great measure of courage. No, acting like an animal to get close to an animal, acting like a plant to get close to an animal, those are things biologists do. But still, the decision to act like a toad is a decision not everyone would make, and furthermore, I would not make except on the sands of some Arthur Bay. I don't crawl along city sidewalks. But I am very alert to that tiger beetle beauty wherever it can be found. And I am very aware that there *are* options, behavioral and cultural options, that will lead me to that beauty regardless of the form that beauty takes.

The elegant experiment, is that not beauty? The clean glassware; the perfect culture medium; the student who follows your best lecture with a most penetrating question; your child making a decision that makes you, in turn, so proud; the wren that actually uses the house *you* built; the title remembered from a hazy past but plucked from a browser's library shelf; are those not also things of beauty? And do those things not also require *the* correct approach, *the* correct frame of mind, *the* purposeful circumstance that brings out not only their beauty, but also your own willingness to see that beauty? Yes, to all those questions. And cannot you remember

Back in Keith County

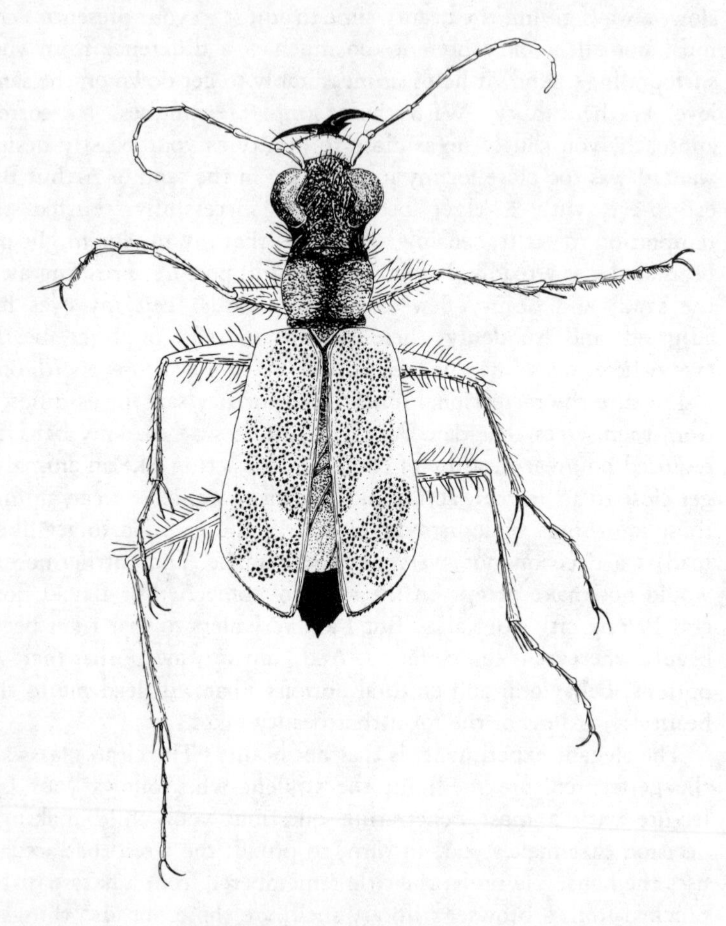

". . . more elusive than the toad, the beauty more elusive than the beast . . ."

times when pursuit of that beauty, maybe beyond the tiger beetle, maybe the beauty of that swallow seemingly beyond you, required some special approach? Of course you remember the time, for you are human, and those opportunities to aspire come many times a day. And you must know that the continuous opportunity to be a toad, to have a tiger, to wish for a swallow, and to feel for those who don't also wish for swallows, that continuous opportunity is your own Keith County.

A Short Explanation

The next chapter concerns rivers. The title of the next chapter was suggested to me by a young lady from Omaha, although she denies it categorically. I distinctly remember walking those river sands with a group of people and I distinctly remember her comments that brought forth the image I was searching for. It had been a frustrating three weeks. My river was in flood stage, and when she's in flood stage she's not good for much. I distinctly remember the day she was again accessible, the day we finally walked out on her sandbars, and I remember that day searching for a framework within which to place my personal feelings about her. That's when the young lady from Omaha suggested the title for the following chapter. I would never have expected such words from such a nice, polite, cultured, intelligent, young lady. But then that young lady also had the grit to ask Waldo Haythorn if she could buy one of his favorite horses. Someone who does that is not afraid to come out with the appropriate images, expressions, that describe a river. Still, it's pretty tough language for a co-ed.

8

The Town Whore

"SHE ain't the pertiest I ever seen," he said, "but she'll give a guy 'bout anythang he wants. She'll go 'bout anywhar, do 'bout anythang, acts lots o' differnt ways, don't cost much. She sure never give me nothin' but pleasure, *pure pleasure,* an' I visit her regular, *real* regular."

Yeah, I thought, and here I am a respectable university professor and *I* visit her *real* regular, too. I mean *real* regular!

"She ain't changed much since back when Ise a kid. She's always looked 'bout the same. Town here don't think much of her. Some folks I'm sure'd like to see her off someplace else. But the kids come down here, you can bet on that! Those boys get a certain age, they're down here in a hurry! Can't never keep the boys away from her!"

Yeah, I thought, if I were a young lad from this town, you couldn't keep me away either! The old man wore railroad overalls, a railroad cap, and he wore like a badge the wear of a life on the cars. He held a fishing pole. He was fishing for his bait. Get to fish twice that way, he said, once for your bait and once for your fish. He was catching chubs and shiners with a very small hook. But he kept an eye on the subject of our conversation, almost as if she might do something unpredictable while his attention was distracted. In that case I was ahead of him. I *knew* she'd do something unpredictable. But I also knew she'd do something *very* predictable, and that was give you everything you wanted and then some, most of it pure pleasure. He kept his eye on her, but I was standing in her. Her name is the South Platte River, and for me—sorry about this lapse into earthiness, folks out West—but for me she's the Ogallala Town Whore.

Ever see cars parked furtively alongside a bridge down by some river, then look quickly as you pass over that bridge and not be able to see people fishing? There is something sinister about that, isn't there? Docsn't it make you think there are some people down on that river that are up to no good? Wouldn't surprise me much if one of those cars was mine. It would surprise me even less if two of those cars were blue vans formerly filled with students. I've seen those cars along bridges, wondered about their drivers, and I've been the driver of cars and vans that ended up parked furtively along a river bridge. So I know about the feelings you get from either perspective. What makes it funny is that no matter how much you know about what's going on under that bridge, you still have the same thoughts when you pass over the South Platte River. You can't help thinking: Wonder what the hell's so interesting down there? It's sort of like maybe there's something going on that you ought to try to be a part of, something with a taste of spice, something like might be going on in a river of ill repute, something some bunch of yahoos would enjoy. You'd be right on all those counts. There *is* something you ought to be a part of, something with a taste of spice. It's called "seining and panning for wild animals in the Ogallala Town Whore."

We have a general rule in our business, and it is this: When all else fails, take 'em on down to the South Platte. You may not always get what you want, but you'll always get something. And it won't cost much. For example, let me tell you what we got this year, the year of the flood: A lesson from a fish. As for the flood, it wasn't really a flood, probably, but more of a special snow in

Colorado that took a long time to melt, or a lot of heavy rain up in the mountains. But when you're counting on her channel to be forty feet wide and calf-deep, and instead it's bank-to-bank and a little over, half a mile of water over your head flowing twenty miles an hour, up into pasture edges, up into trees, carrying big logs on down to the Missouri, then that's what I'd call a flood. It's sort of like coming on down to the River of Ill Repute with a little seine and a lot of big expectations only to have some mature and unruly Madame say, "Sorry, kid, y'aint got what it takes today." And there you are standing up off the bridge with your little seine thinking, "Wonder what I'm going to do instead?" It's kind of as if you hadn't planned to be told, "Y'ain't got what it takes today, Sonny." Well, it's still only the Old South Platte. While you may not have what it takes to get what you want, you've still got what it takes to get something, that is if you have the courage to cowboy-it right on in.

In my life of late, it seems the saga of the plains killifish goes on and on. I've written more about the plains killifish than about almost any other single animal: grant proposals, requests for money, scientific manuscripts, parts of scientific manuscripts, it goes on and on. I never set out to be an ichthyologist, or even to study parasites of fish. But from that first visit to the Town Putout, the plains killifish has taken me like some addictive drug. There is still not very much known about the fish. A lot of people can describe exactly how and where to catch them sometimes, but not many can tell you exactly what a week-old killifish eats, how often an individual must eat in nature, at what time in their little wilderness lives they acquire their parasites, exactly where they go and what they do in the winter, or what kind of special mysterious relationship they have with the South Platte that weds the two species—the living river and the living fish—together forever and forever, through sickness and in health, till death do them part. So now I wonder about my own personality, what there is in it that would involve me in this saga of the killifish—a saga that takes me out on the highways in unruly vehicles, that finds me in the Platte in February stomping ice to make a seine hole—exactly what it is that would make me feel a need to try to become the third point in a love triangle involving the Town Whore and her septic companion, *Fundulus kansae,* the plains killifish.

Interesting pair, that couple, I thought, standing on her banks looking out across a flood that could not disappear in two weeks, that was half a mile of raging water sweeping away the plains into the ocean. No way to keep your feet in *that* river, at least this month, I thought, standing there with my little seine and play bucket. Those might have been the thoughts, but when you've driven eight miles to catch a fish the size of your finger, you don't turn back just because of a rampaging flood.

"Aw, hell," I said, "let's just go in there and seine up some o' them babies."

Of course she agreed and picked up her bucket. She was the frailest girl to have ever challenged the Platte.

There is no better way to understand the structural complexity of a braided prairie river than to try to seine fish during the floods. In normal times you take the river for granted, assuming there will always be those microhabitats wherein wavers the killifish. In normal times you can stand on the sand and look up river into the explosion of setting sun and see a hundred, a thousand, places perfect for the killifish. In normal times you can walk out there right before daybreak and see them moving, gently holding their places, maybe riding the current just to be riding the current, watching just to be watching, maybe just enjoying the special time before the killifish workday starts—all in normal times.

But in flood times it's different. The water's muddy, it snaps at your feet even as you stand dry on the gravel. It feels hard when you step out into it, and small logs, cowpies, beer cans, sheets of plastic, all loom out of the distance and rush between your legs, piling up under the bridge in dark whirlpools—all at flood times. It gets deep in a hurry and your feet go out from under—in the flood times. But as you stand in that torrent, cold and dirty beyond your dreams, the smell of feedlot creeping up your jeans along with the water, you smile. For looking out across the flood, it is obvious there are more places to be than a person could count. And in one of those places is a killifish.

There was the place with all the small cottonwoods, flooded and maybe because of it dying, that day, and we seined the trees but there were no fish. There was a small spit of sand out across the torrent, so we waded to our necks to reach that spit where alongside there should be shallow water. We seined the sand but there were no fish. There was a place up under the bridge alongside a piling, where maybe a killifish could find a refuge; but then I decided maybe the killifish didn't want or need a refuge, for we seined the piling and there were no fish. Now in normal times one gets a distinct feeling of compartmentalization, walking the Platte. Although the killifish may skitter away, you never have the feeling they will stay away. No, there is always this sense of *a* kind of place for the fish: a perfect combination of shallowness, warmth, current. You never get the sense that the perfect combinations are smeared together with the imperfect. There is a feeling you get, in normal times, walking through river parts that look exactly alike, that when you take a certain step you've moved from the imperfect to the perfect. There is a sense of having entered the killifish's compartment. Does spending so much time in the river turn a person's skin into that of a fish? Are there scales, lateral lines, mucous, on the legs of a person who wades the Platte so often, and do those things allow one to feel the river as does a fish? Maybe. I stopped looking for the killifish compartment and began feeling for it. We waded beneath the bridge and downstream, alongside a plain bank with large gravel, rocks, and I took that one step when the feeling came.

"Let's seine here," I said to the frailest girl. So we seined the rocks and there were killifish. Not many, but some; enough.

Usually when one seines the river, there are many kinds of fish to be thrown away: chubs and shiners, carp or sucker fry. But on that day there were only killifish. And days later, on another part of the river, we again stood on the bank, our minds doing what the killifish would do: seek that portion of the flood just right for *Fundulus*. So again we waded into the raging South Platte to find only the killifish; again not many, but some; enough. It was only many days, even weeks, later, when the South Platte began to go down, when sand bars began to appear, when beer cans stuck in the mud on those sandbars, that we found the others: crayfish, carp, a few chubs, shiners again appearing in the nets. No professional biologist would ever say those other species disappeared. If this love affair were with the creek chub, then perhaps all I'd said above would be about the chub. Perhaps it would be the chub to which our sense were tuned, or to which we thought our senses were tuned. Perhaps we'd be able to find those chub places, where the chub hides when the Town Whore goes on a rampage. Perhaps; but I'm not so sure. I never felt the killifish was hiding.

No, I've had the feeling since the flood that my killifish is not exactly the animal I thought it was five years ago. There is another side to this personality. This other side leads my fish to run with the unruly Lady, to stick with her when the others scatter, to tolerate her rampages, to stay through thick and thin, through sickness and in health, to be unawed by her excesses, to take all advantage of her in her gentle and vulnerable times, to share her with others, to be true "in my fashion" until death do them part. And you must know that indeed death will them part, the Town Whore and Mr. *F. kansae*. There is no better demonstration of that than at the city of North Platte, where the river is temporarily killed. The Tri-County Irrigation Dam is located at North Platte. The Town Whore backs up behind this dam; the Town Whore is sent out down the Tri-County Supply Canal into domestic duty, agriculture, the watering of fields; the Town Whore is stripped of her sand, dredged continuously; she is made respectable or so it would seem. And so it apparently also seems to the killifish, for immediately downstream from that dam there are no *Fundulus*.

I am very certain that in some, perhaps a great many, offices and homes there is a feeling of satisfaction that the Town Whore has been made respectable, put to good use, domesticated. I am very certain that in many offices there are professional people who are proud of their accomplishments, domestication of the Whore. I am totally certain that the watered fields along the Platte contribute significantly to my own economic well-being. And as certain as I am of this last fact, I am just as certain that I will never be allowed to forget the extent to which those watered fields feed my mouth. I am also just as certain that all across my land, in place after place, there are wild Whores being put to domestic work by smug architects of progress who look out over their work, feet up on some government-issue gray metal desk, and are satisfied, secure, content, accomplished, that they've "reclaimed" some Whore. They are cuckolds, these men. For if given half a chance, the wildness will return. These men do not *really* know the relationship between the killifish and the river. That's the reason they are cuckolds. Thirty miles downstream from Tri-County, the plains killifish can again be seined.

There is a disrespect for limits in the Platte's unfaithfulness to the men who would tame her. There is a celebration of option in the flooding Platte's obliteration of those sandbar edges I find so intriguing, so essential to the killifish fry, so appealing that I can often for days think of almost nothing else. There is a scorn of history in her burial of special channels where I have gone for years to get the chub, the shiner, the white sucker, and the snail. There is a vixen in her flood. That vixen tells me in the most haughty of icy fire that my plans for seining the South Platte that day mean nothing to her, that it's time to use my teaching skills to find another experience for my students. But there's a touch of sentimentality in the detritus of her wrath: little blue bottles with strange writing stick in the mud after the water goes down. I find things that may be valuable antiques strung out along the sandbars that appear one day out of the morning sunshine. But most importantly, when her water goes down, in those days right after the flood, she is fertile beyond words. She traps millions of crawdads in drying pools and you must know there are many raccoon tracks across the mud, and many heron tracks, in those places where crawdads are trapped. There is a smell to her after a flood, an organic smell, of dead fish, rotting plants, mud laden with decomposure of a thousand feedlots. But somehow that smell is a good smell, when it's out on the South Platte. But most of all she leaves behind a lesson in the protean. If there is anything you learn from a trip to the Town Whore, it is the manner in which one deals with a changing time, a changing place, a history and geography with no respect for history and geography.

Throughout early American history the Platte was a river of opportunity, an avenue of exploration, a road across the high plains to mountains, on the way to a fertile West Coast. Those early pioneers followed the Platte for hundreds of miles, constructing a windlass to pull their wagons up a famous hill, but knowing all the time the Platte was a river of life. The teeming wildlife beckoned, seemed inexhaustible. They trapped and ate that wildlife, and sold skins, but some eventually settled along the river and knew they must also grow food. They turned to the river for water. Although they loved their river in a way only those who've lived together through the hardest times can love, they nevertheless began trapping the river as they had earlier trapped the mink. And as before when they viewed the wild things as an inexhaustible resource, they also viewed the river as an inexhaustible resource, not realizing that it, too, like the beaver, the mink, and cranes, was a wild thing.

You can see footprints of pioneers on the Platte today, but they're a different kind of footprint and a different kind of pioneer. Little boys growing up along the Platte explore its braided stretches, wade after things in the water that comes out of the mountains. But sometimes, it seems, when those modern pioneers get grown up, they get themselves elected to the charged atmospheres of legislative chambers, they get themselves appointed to state agencies, and the old Platte they knew as kids is forgotten. It seems, sometimes, that in their minds the rampaging Whore, the Sentimental Vixen, the meandering channel, the sand, the tiny blowing cottonwood trees, the snails and running plovers no longer exist. There remains in their minds only the water. If they remember the killifish, they don't admit it. I've never read the words *"Fundulus kansae"* in any newspaper report of actions concerning the Platte. No person of power, no Governor, no Public Power and Irrigation Official, no football coach, no owner of any newspaper, has ever publicly noted that the plains killifish is absent from the Platte for many miles downstream from the Tri-County diversion.

But the briefest of lessons in the protean, if remembered, will engrave that relationship between the Platte and a fish into a mind forever. How, one could ask, can such a strikingly colored, delicate, little fish assume such importance in a mind that it cannot be forgotten? That question is easy to answer: *F. kansae* does not survive well in captivity. The species evidently *requires* the Town Whore and her uninhibited moodiness—bone dry at times; raging with floods out of the mountains at others; frozen into jagged glaze in winter; warm, clean, and gentle in early fall; drying sandbars with nurseries for killifish eggs in June. Wade after killifish in the South Platte for a few years, and you come to think of the river only in terms of its *Fundulus*. You wonder where they go in winter, how many embryos get swept away to the ocean. And when you think of the river in terms of the killifish, you are thinking of that river in terms of a little striped fish that has no economic value, is not even good bait, can't live well in an aquarium, and is not at all endangered, at least as the term is officially defined.

Now so often, it seems, the scientist feels a need to explain to the public the value of some nondescript little creature with no apparent economic value and whose continued existence blocks "progress." The scientist so often appears a strange species that would deny the electricity, power, money, water, and attendant good life, to other humans he doesn't even know. He is asked to defend the life of some apparently valueless creature in terms of the economics that we all only think we understand. At those times, so often it seems the scientist fails, or at least compromises some deepest inner feelings about the value of wilderness, and then considers the compromise a victory. In this nation of mine, it often seems the lives of obscure creatures have so little importance when weighed against the accelerating technological condition, the political struggle with laws of thermodynamics, the ultimate effects of DNA replication, the modern sociobiological manifestations of primate evolutionary history. How many times I have stood in the Town Whore, sunrise exploding through the cottonwoods, and thought how interesting it would be if the politicians had to justify the demise of some creature like the killifish, and had to express that justification without once mentioning dollars, business, or human food.

Could they do it? Of course not. For when the individual human thinks of the collective welfare in terms other than dollars, business, or a mouthful of food, then the individual places value on things that allow long-term human survival in the *human* condition. *All* animals vie for food, compete for shelter, kill for economics, fight for energy, stake out territories. Only the human could look at the plains killifish in a world of accelerating change, in which stability seems extinct, and conclude that the killifish could teach us a lesson about how best to survive in that ever-changing environment. Only a human could see in the killifish and the Platte a relationship always true through flood and drought, feast and famine, the turbulence, meandering, unfaithfulness, shallowness, undercutting, uninhibited whoring times. Suddenly, I think, a species of fish that could help us learn how to survive those kinds of times might have some economic value after all.

That's probably why I find myself down on the Platte so much. Every day, almost, my dreams include the killifish and its river, but maybe those dreams are only reinforcements of real times. We seined Minden one time, in February, and a kid from Louisiana came along. We didn't eat for hours and hours and after that trip all he could talk about was how we didn't stop to eat. Minden was the place where the seine froze solid when we tried to roll it. Then we went to Ogallala in February, after we'd seined Minden. When we woke up the next morning there were five bald eagles sitting in one tree. We seined Maxwell. It was a hell of a lot of fun seining Maxwell but there were no fish. Grand Island. We tried to seine Grand Island but went to Phillips instead. Phillips was kind of fun; we had Amy along with us on that trip. I remember finding the most beautiful freshly killed meadowlark on the highway, bringing it home, painting its picture. The melting snow made interesting patterns in the gulleys at Phillips. One of my white buckets sitting in the corner of my city lab has "Silver Creek" written on it in grease pencil. I think at Silver Creek we seined the city dump. Kearney, that was an interesting trip. Had to wait in a motel coffee shop for it to get light enough to seine that trip. I like to start the day early. I took a beautiful picture that day near Kearney; it shows Anne and Jim popping the seine with this shower of droplets, all in the yellow light and shadows of daybreak. We seined Tri-County; no fish. Then we seined North Platte; millions of fish, all infected. So you can see how my daydreaming goes. Oh, I forgot about the ice. We found out you could run your seine back up under the ice and get some killifish. But my waders had a leak so I got up on the ice and of course it broke. I remember kicking that big chunk of ice on down the river and seining some more killifish right in that same spot. These are all the kinds of thoughts that occupy what should be "productive" time. It's almost as if I had some chronic disease that reduced my contributions to society. That's it, a disease. You know how in some tropical countries disease can wreak havoc with the productivity and social structure? That's about how it is when I start daydreaming about the Platte and the killifish. Disease, that's what I've got, a social disease. Messed around with the Town Whore for five or six years and came away with a social disease.

I'm dreaming now, in the winter, of the Platte in June, and wondering what her mood will be then. But there is one thing I do not need to wonder of and that is of her friend the killifish. He will be there, surviving her floods, savoring her finer times, in all those places, of course, except Tri-County. Mr. *F. kansae,* if questioned, would smile at the obtuseness of those who would question his relationship with his lady, of those who could not understand the lesson to be learned from one who would associate only with the wild, of one who would shun the domestic. Late at night this winter I searched for pages to help me understand my land, my river, and the times in which I find myself. I found those pages in Barbara Tuchman's *Distant Mirror,* in the Durants' *Lessons of History.* In the company of those pages the South Platte became the river of mankind's time on earth, and vice versa, with floods and droughts, ever changing but never changing opportunities, moods, shallows, and holes. How can it all seem to change continuously yet seem somehow the same? How can we seem to be fighting those same battles over and over again, but in different places, with different opponents, so that there is always this sense of having been there before? How can we always seem to see new things we want, but not be able to make that connection between a desire and a reason, not be able to tell ourselves *why* we want something? In the company of those pages, and in the company of those questions, I turn to the Platte and the killifish. I find one that is wild, unpredictable, knows no limits, can't really be tamed very well, sentimental, turbulent, meandering, unfaithful, and uninhibited, and I find another who *requires* that protean richness. When that happens, problems become opportunities. When problems become opportunities all because of a lesson from a fish, then I find it difficult to see how this species of fish has no economic value beyond that of food for kingfishers. Maybe the value of a two-inch fish that could teach humanity how to thrive in whoring times is so great that we can't really recognize that economic value. Maybe the same thing might be said for all sorts of obscure little wild things.

My sense of the value of *F. kansae* comes, of course, from a study of the way this creature deals with an unpredictable environment of which it can only be said that such environment *will* change. When unpredictability is all that can be predicted, the critical element for survival must be the unhindered choice. The killifish must always have the option to swim anywhere, the freedom of total choice. I am still enough of a biologist, however, not to romanticize too much this matter of unhindered killifish freedom. Wild things have their limits of tolerance. They are kept within those limits by genetic instructions. But I also know that nobody, but nobody, tells a wild thing where to go or where not to go in search of critical elements of survival. And from my experiments I know, finally, that the one thing the plains killifish will not tolerate well is an environment provided by someone other than the Platte. I know that survival in the Platte requires unhindered freedom of choice and I know that because I've learned it from a fish.

9

Lido

THERE was a girl who came to work in my laboratory once, with very high grades, from out of the north land, wearing long red hair, green eyes, and freckles. When she left, she left with a Ph.D., a boyfriend, and a boat, and headed back to Minneapolis for a few years in medical school. She bought the boat in Omaha; I don't know where she picked up the boyfriend, maybe from among her sailing friends. Come to think of it, she might have fallen in amongst those sailing friends out of frustration with the local culture. After all, Nebraska is not quite the same as Minneapolis, at least that's what everyone from Minneapolis tells me. Well, if her sailing friends here are anything like people in Minneapolis, then they're a bunch of real competitive souls up there in the north country, for about all she did with her sailing friends around here was race. Oh, she did a lot of talking about racing, too, and about cats turtled out in the algae, broken rudders, and beer. Through all this conversation ran a paracommunication of intensity, of human striving against human, of glory in a broken mast only if the disaster happened in the heat of some contest of wills. Tales of gore and splintered steel cables and shredded Dacron and sweet victory with Lite came every Monday during "the season" after every

Sunday afternoon of "the season." And on every Monday evening, after such tales, I would go home, walk into my garage, and talk to my Lido.

"I don't really think a sailboat was made to race; do you? Is that really your purpose, to extend this Medieval joust we humans call a workaday world out into the winds of the Third Planet?"

And every Monday the lines of the Lido would reassure me. I would squat down on my heels, the better to see that design along the bow carved in fiberglass, that curve of the gun'l, and every Monday my blue mistress would answer, "No, it was never intended that Man should race a sailing boat. Man races other, lesser, machines; Man races other Men. But a sailing boat was given to Man so that he might tether himself to the wind. That is a higher purpose, to be tethered to the wind." Then every Monday we'd stand together for a while out in the garage, my hand on her deck, not talking anymore, just watching the leaves blow out on the driveway, and thinking about every wave we'd ever hit together, every cat's-paw we'd ever caught, and about all those times we'd spent those blistering hours doing nothing but riding the wind just for the sake of riding the wind. Nothing more, just the wind, a Lido, and some guy who wanted to be with the wind for a few hours, all standing remembering each wave, that's what it would be like on those Mondays out in the garage.

I *do* have this love affair going with the Lido. This is a luxury upon which I spend great amounts of intellectual energy. It has all the elements of a true love affair: the immunity to outside influences, the times of infatuation, the times of frustration, the reluctance to abuse, the strained patience toward others, the reassurances of a mast hanging along the garage ceiling as snow blows against the doors, and the little surprises that creep into a life when we're apart. Then there's the bit about the lines, the watching and staring from every possible angle. It's parked in the garage in the summer; every day I walk through the garage and look at those lines from a different angle. Sometimes I just stand there in my city clothes at the end of the day for a long time with my hand on the deck. Sometimes when I do that I can feel the burn of the mainsheet cinched around a wet hand straining against the wind.

Now *there* is an element of nature that all prairie creatures know: the wind. The wind comes out of The Dakotas in winter, carrying a harsh cutting bitterness you must stand and feel to understand. The wind comes out of Kansas and Oklahoma in spring, carrying an organic smell that means danger for anyone dependent upon responsible behavior from people like me. That Kansas wind is warm and moist; it makes you smile before daybreak and it blows back the swallows. The wind shifts suddenly, but by then you've usually been warned by the thunderheads or by the sheets of rain carried on the wind to where you can sit watching from a bluff. The wind also goes away, leaving a lake of total smoothness, and although they're so far away you can't see, you can nevertheless hear them: fishermen talking in the stillness way down the lake, a duck quacking in the stillness way down the lake. Everyone out here knows all *those* things about the wind. But a few know some other things about the wind, such as the fact that the wind blows away my mind, too, for I've learned the movement of every leaf of every tree I can see from my city work building. When those leaves move just right, my mind goes home, hitches up the Lido, and heads out across the prairies for some stretch of water and some lee rails down in green-algae green prairie water.

People in my profession think the Lido is a luxury. I'd like to put that thought into some kind of perspective now: the man across the street is an electrician, he has several-thousand-dollars-worth of boat parked in his garage; the man next door is a milkman, he has several-thousand-dollars-worth of boat; the man who used to live across my back fence was a truck salesman, he had a self-contained RV mobile home parked alongside his driveway. I am a university professor, so what the hell's wrong with a university professor having a boat? Nothing, of course. That was my thought one day when I was thirty-eight years old and decided, having lived seventeen years since my last sailboat, that I could not then live another minute without the next.

The Lido cost $2000. It had sat on the T & R Yacht Company's lot for a year, unsold, and was slightly damaged. It was love at first sight. That was nearly five years ago, and the Lido has paid for itself in noneconomic terms a trillion times over. "How can you afford *not* to sail every day?" people often ask. The answer to that question seems so obvious: there have been times out on Lake McConaughy, the largest lake of Cowboy Country, when a single day was worth the entire purchase price of my Lido. Knowing what I know now, if some day in the future I was guaranteed such a day out on Big Mac, and did not have a Lido, I would easily pay $2000 for that day in communication with the wind. Don't recoil in disbelief. Think about it a little bit. Power, money, politics, your friends' divorce, car repairs, gasoline prices, stress, your dog biting a visitor, taxes, hostages, Afghanistan, war talk on the horizon, OPEC, is there more I should mention? Leave it all. That stuff is all the baser human animal instincts at play. Leave it all; come tie yourself to the wind; communicate directly with the planet that supports you; merge your senses into an entity of wind, water, and boat. Forget Dacron and fiberglass, forget they are man-made materials, for although they are man-made *materials,* it is their *shape* that is some designer's successful idea of what it takes to tie yourself to the wind!

Thirteen feet, six inches long, but broad, roomy enough for four adults, since the molded seats go all the way back to the transom and the side decks are but a few inches wide. Blue above, with blue waterline stripe on a white hull. There is a tray at the fore end of the cockpit, a tray to hold the corks that fill the holes in the ends of hollow seats, a tray to hold the halyard wax, some Nosekote, a tray that a generation ago would have held a "church key." Back then beer cans had to be opened with a "church key." The spars are gold extruded aluminum, but the gooseneck is cast aluminum and breaks easily if the mast is lowered on it. Stay and shrouds are stainless cable sheathed in polymer, sheets are nylon, blocks are nylon and steel, travel is adjustable, and the foot of the mains'l can be loosened to spill the wind, or to add curvature, but when you spill the wind from the main, that wind spills out on your feet.

There was a boom vang; I took it off. There were hiking straps; I took them off. Vangs and straps are the accoutrements of racers. A clean cockpit is the accoutrement of a man who would be at one with the wind.

Jib and main are Dacron. Jib sheets run outside the shrouds and I always, *always,* give a guest the responsibility for jib sheets. It makes them feel a part of the operation, a part of this being at one with the wind, with the planet. It also keeps them from sitting on the jib sheets at critical times. The jib runs up the stay, the main runs on a track in the mast. Into the leading edge of the main is sewn a heavy cord that runs in the track. Battens are fiberglass and the longest is the middle one of three. Halyards are steel, stainless cable, with nylon spliced onto an eye. The trailer is angle iron painted white, nontilt, with standard tires, one-inch axles, and wheels with Tyson bearings. Wheel bearings are packed with Sears best grease and Janovy's best loving care. The trailer had two welded tie-down loops that have long since broken off and have been replaced by loops bolted through holes I drilled myself in the angle iron.

The rudder hinges through a pin at the top, and is wood covered with fiberglass hinged into a brass fitting. When the fitting is in place and you're far enough out, you can put the rudder down and clamp it with a big levered bolt through this fitting. The tiller is about three feet long and had an extension on it. That extension went the way of the vang and straps; tiller extensions are also the equipments of racers. The centerboard is hinged forward, and thus the angle, and center of lateral pivot, can differ, depending on how far down you put the centerboard. I used the vang hardware to rig a lock for the centerboard. The lock consists of lines running through opposing jam cleats. The vang business was a nuisance; the centerboard lock is essential. As in any simple trailered day sailer, when the centerboard and rudder go down, then there is a bite, a catching hold, a surge forward, lines and cables go tight, you can feel all that in the seat of your cutoffs, and it's a feeling that tells you the connection between wind and earth has been made. When you get that feeling in the seat of your cutoffs, then you need to know that centerboard is locked where you want it.

That bite, that catching, must be the same sensation that astronauts have when the docking maneuver is made, when contact between two elements of the universe is made properly. And as in the case of astronauts, that electricity of connection between vast natural elements is made through a human body. You are a conductor; communication with the wind flows down the length of Lido main, out onto the mainsheet, through your arm, through your rear end, through fiberglass, and into the depths of the water; that's what that first bite, that first catch, feels like.

Cast off technique is summer special at Big Mac: there are no deep water ports, at least for public use. Oh, the power boats pull up to those floating docks that run out alongside the ramps at Martin Bay, but power boats are power boats. No sooner do they run up against the dock than they pick up some passengers and go a hundred yards down the beach where they run right up on the sand and unload. Besides, the power boats don't draw as much as the Lido with her centerboard and rudder down. For the summer special you wrestle the Lido off the trailer and wade it around the power boats and pull it up on the beach to step the mast. Then you have to wade out far enough to put on the rudder, put the centerboard down, raise the sails, etc., all with someone paddling out. There must be a hundred more seamanly techniques for getting underway other than the summer special. I'm sure there are thousands of people out there, accomplished in long years of seamanship, who would without thinking handle getting underway very differently. There would be sails raised while the thing was still beached, there would be rudders on while that blue waterline rested on hot sand, there would be spars and sheets rattling in the wind, the whole thing would appear smooth, a merging of technique and human desires. But you must see where my approach to getting underway comes from; you must see wherein lies the summer special, the reluctance to raise sail except in deep water, the paranoid care of rudder, all that. It comes as part of the affair. I have this sense of exactly what stresses and strains my Lido should be subjected to. It may not be the correct sense, but it is the sense that dictates our relationship on the beach. There are some things you just don't do with women, things like walking along the inside on a city street. As so there are some things you just don't do with a sailing craft, things like stress them in ways you don't *think* they should be stressed. Who knows if all those special things are necessary; who knows? No one really knows if senses of treatment are at all necessary. All one sees is the results. The Lido has never been damaged. Nor have I; the relationship has been nothing but freedom and pleasure.

R. D. Schock, Newport Beach, California, the manufacturer; 3962 the serial number, and 3962 the blue numbers raised against the blue cowboy country sky meeting the blue Big Mac waters too far to see across a blue deck right above a blue cooler with Pabst Blue Ribbon all sure to leave behind in shambles any vestige of the blues brought on by whatever-ails-you. I've taken to raising sail myself, turning the tiller over to any intelligent crew. It's easier to say "push" or "pull" than it is to judge the trim of a sail. It's easier to call out "push!" or "pull!" than it is to adjust the downhaul, the outhaul, or to let out an inch on some sheet, take up an inch in some halyard.

The Lido has an interesting design feature that is not a part of every sailing craft. Trailered, the golden aluminum mast lies along the top of the boat, resting in blocks and tied down with the running rigging. Out on the interstate, I tie a red cloth to the halyard loop at the top of the mast. The fixed rigging, except for the stay, remains attached, with the shrouds running down through holes in the deck to pins (adjustable) on the front of the seats. To step the mast, you untie all that running rigging, carry the mast aft along the deck, fit it into the step, then bolt it in. You then stand pulling on the stay while someone raises the mast, then tosses it into the air, after which you catch it with the stay and pull it all the way up. The unique Lido design feature is this: When the boat is pulled up on the beach, on that two-thousand-dollar-day, there is some unique combination of lines, shapes, proportions, so that when that mast goes up there is a tremendous flood of emotion. You almost shudder with it. No one talks, for with that act of raising the mast, you've entered that realm of freedom known only to other creatures that ride the winds. I never get that feeling watching other sailors raise *their* masts. I hardly ever get the feeling they are there only to be at one with the prairie wind. I get the feeling they are there to race.

There are probably a thousand bays like Martin Bay. There are probably a thousand places where you have to get out of, away from, for those thousand places probably reek of humanity as does Martin Bay. Right at the mouth, that's the only place Martin Bay doesn't reek of humanity. The parking places disappear, at the mouth, and the road runs out into deep sandy gulleys between gigantic barren dunes. Sometimes you can see people way down that beach below the dunes. You almost always think those are people who want to get away from the crowd; there isn't any other reason to go down that far toward the mouth of Martin Bay. Across the bay from the dunes there is an island, on the west. Since the wind is almost always from the south, you have to tack to get out of Martin Bay. But the trees on the island bend the wind a little bit, so it's not really a difficult tack. Probably the most difficult part of the tack is seeing that beach with the dunes and the people who don't want to be with the rest of the crowd. That's the place where you're suddenly out on the main lake. You can't get to that place fast enough.

It's total impatience tacking out of Martin Bay, watching those people, seeing them watch you, waving, just waving at some other people *they* know want to get away. Sometimes those people have little kids with them, sometimes dogs. Sometimes when they have kids and dogs, the kids and dogs run along the beach waving. No one else waves, except Karen out sunning, seeing those blue numbers, 3962, out across the sunstruck waters of Martin Bay. But the people on the dunes at the mouth, they almost always wave. They're kindred souls, probably, or they wouldn't be where they are. But then you look back and those people are still waving, getting smaller, and out ahead across the miles are the bluffs of the south shore, and in your hand is the mainsheet, and it's cinched hard around your hand, and through it you can feel the wind, so you sit up on the little deck and put your sunburned foot up on the centerboard well. And the wind takes hold, and you look for a while at the sails, and it all goes away, all that's ever been a part of some vegetative responsibility, some sets of rules, regulations, obligations, tasks, races and more races, fights and more fights. It all goes away, and it's just you and the boat and the wind and the intelligent crew and that mainsheet, no don't forget the mainsheet, cinched around your hand. On the other end of that nylon rope is the prairie wind. And though they're so far gone over the horizon you can't see them, somehow you feel that those people on the dunes at Martin Bay are still standing, waving at those blue numbers, 3962, and themselves wondering how to hitch a rope to the wind.

Sliding air masses, the jet stream, cold fronts, warm fronts, warm moist air from the Gulf meeting cool dry air out of Canada, still air, hanging-in-shrouds air, deceptive air, Branched Oak air, thermals, downdrafts, puffs and cat's-paws, to wind'ard and to lee'ard, but a mile above the masthead floats a redtail hawk. The hawk turns into the sun, but the sun turns and runs down the gold mast, casting high value shadows over on the main, the one with 3962. The hawk turns on a long glide into the haze above south shore bluffs, but the south shore bluffs turn, moving lazily beneath the jib, then out beneath the boom and off away to Lewellen. The hawk is gone into that haze over the bluffs, but there remains a mile above the masthead the essence of hawk, that essence of redtail that tells of a union with the wind. Somewhere among those long feathers, out along that wing, and on that rump where rectrices become body, there are the sensory receptors for sliding air masses, the jet stream, cold fronts, warm fronts, warm moist air from the Gulf, still air, hanging-in-shrouds air, deceptive air, Branched Oak air, thermals, downdrafts, puffs, and cat's-paws. And somewhere out there beneath that hawk I've crossed the 102nd meridian, tethered to the wind by a three-eighths-inch nylon mainsheet, wet, sunburned, pulled into the west and into those flights of fancy that only a sail can evoke: pirates, simpler times when men did not know of engines, simpler times when it seemed the natural thing to tether yourself to a wind and ride that wind into history as an explorer.

Someone said the Lido was designed for saltwater day sailing, with its broad beam, stability, and its bite. I don't know whether that's true or not, but what I do know is of the bite, or at least what I call the bite. To watch it is to see another world of design, another world of movement, another world where there's only the water, white against a white hull, clear yellow with refracted sun, black into the depths with fleeting impressions of large fishes, and the white fiberglass reflecting all those colors as the designer's lines bite into waves. The Lido never swerves, never slaps back at waves, is never pounded aside underway, never intimidated by water; no, the Lido bites, cuts, welcomes, almost, the challenge of a wave out over the 102nd meridian, and commands the water, carrying me off on some mission into the unknown reaches of the mind freed from races humans race.

There is a feel to this boat, a feel of solidity, of a need for steady wind and long reaches, for that sandy bluff seen below the jib. When you reach that bluff, you'll feel a million miles from everywhere because you sailed there. It's that feel of having gone straight and long away, knowing there's somewhere at the end, something at the end, but not knowing where or what. That's the feel some designer put into the hull of a Lido. How many times we've been to that somewhere together, and how many times we've come back having found that something! That somewhere is the place you don't have to go any farther to know you've put into perspective some race run back in the trenches. And that something is the knowledge that you've been there, dragged on the prairie wind, carried by a hull that bites the waves and never swerves. And I guess the knowledge that we've been there is what makes me stand out in the garage sometimes, dressed in city clothes and carrying city thoughts, and put my hand on fiberglass to feel those lines that feel best crossing the 102nd meridian.

Wings of Dacron, an imaginary line from someone's brain carved into fiberglass, golden aluminum to release that flood of emotion, and a nylon rope to wrap around your wrist, hitch to a sliding air mass, and make yourself at one with the redtail; those are all the things you can buy for a couple thousand dollars American currency. Wings, are not wings the symbol for freedom? Are wings not the things with which creatures catch the prairie wind and ride off into their destinies? Are wings not the things with which creatures can turn, choose, ride a hot thermal off to another place in a straight line? And are wings not things that give lesser animals powers no human could dream of having? Yes; yes to all those questions, and yes to still another question: Do you also have wings? Of course; they're made of Dacron with big blue numbers, 3962, and if you think for a moment I can't put them on and ride a hot thermal off to another place in a straight line, then you are very wrong. What seems a luxury, when put in those terms, becomes instead a necessity. I knew it would eventually come to this; I knew even as a child that someday I would have to have wings. What I didn't know then was the extent to which a Lido of the mind would serve as well as a Lido in the garage, and almost as well as a Lido out on those host sands of the very biggest lake in Keith County.

10

Wings

SOME are called altricial, and these are the ones that require much parental care, much warmth, and feeding, the performance of tasks humans cannot do. Some are called precocial, and these are the ones that open their eyes and go out into the world running, darting, hiding, their way into adolescence and adulthood. Some are as ugly as anyone can imagine; some are so beautiful as to release a flood of emotions just by seeing them. Some live in the filthiest of places; some live where it's clean. There is no correlation between beauty and cleanliness; there is no correlation between nestling ugliness and adult beauty, but that you know from childhood stories. Some have patterns that help them escape their enemies by hiding. Some have only the behavioral pattern of studied helplessness. Some you can befriend, take home, raise for your own, and they will love and recognize you for it. Others will be killed, physically and in spirit, by such an act of good intentions. Out in nature most die before they get a chance to reproduce. They find their ways into the bellies of snakes, skunks, raccoons, large bass, or they die unseen out in the grass and are reduced to dust by the scavenger insects. The vast majority intend to fly if they ever can. They move their wings in mock flight before they can see the next branch. It is not encouraged that you learn all these impressions for yourself these days. Nowadays most everybody says to leave baby birds alone.

Starlings. I started with starlings, for no one says to leave a baby starling alone. Starlings are unprotected, killed on purpose in vast numbers, reviled always, or at best tolerated, and there are rumors they carry disease. I was studying malaria at the time and in Machiavellian practicality chose the starling for all the above reasons. I needed experimental animals. It gives me the shudders now to think back on those days of robbing starling nests, hanging upside down from a rotten limb forty feet high, chopping a hole in that limb to get out the nestlings. But from those experiences comes my advice to those who would seek housepets among the wild: begin with the starling. Starlings are more intelligent than several breeds of popular dog and certainly more entertaining than cats. And when the squawking is all over, they'll lift on starling wings to make their keepers proud.

There were times later in those experiments when I'd let them loose around the room on purpose, watch them fly among the ceiling pipes, and admire their touches upon the air. Wild things carry something with them into captivity. Sometimes that something kills them; sometimes it is the very thing that allows survival in a cage; sometimes a human can see that something. That something is what a wild thing cannot shed, like a touch upon the air. Born with that touch, you say, are those wild things not *born* that way? Of course they are, and of course they will all exercise that birthright with the first chance of freedom. And just as of course, starlings are not the only creatures that enter their freedom-volumes with their own unique touches upon the air.

But I've wondered many times in the years since if I would ever have begun to watch for a touch upon the air had I not let those starlings fly around the ceiling pipes on purpose, had I not given a tiny taste of freedom to a reviled disease-ridden import. I stood then, in a supposedly aseptic "animal house" and violated all rules of good sense by letting my experiments fly among the pipes just to watch, knowing at the time a "touch upon the air" was not something that could be quantified. And so I stand now, in the most septic of worlds (Would some call our present condition an "animal house"?) and watch wild things, looking always, because of the starling experience, for some unique touch upon the air.

For some reason, those that fly their own courses are the ones I watch. The fall geese are impressive, high into the evening sky right over the tops of town, above my neighborhood where everything stops to watch the geese until they're only a thready line at the horizon. But no goose flies alone these times of year. Cedar waxwings come to town in late fall, *tswee*ing plaintiff among the berried trees, slamming into panes to be picked up by students who marvel at their beauty close-up. Their texture, held in the hand for a first time, can hypnotize. Our son, resigning his way to school one day, discovered his first waxwing dead beside a window. He hid the bird carefully beneath a bush, then picked it up again on his way home, presenting it to me with that electric tone of wonder in his voice. "Cedar waxwing," was the answer. "They always seem to fly into things and get killed." Then I told him about the flocks. No waxwing comes alone into town. Their communal sound slips through my environment for a few days before winter, then leaves. We talked about winter goldfinches after that, right up to the deck doors after seedy bits, never alone. And he brought up the cliff swallows we'd see next summer, then we finished it off, as the light faded and someone brought in firewood, with some rememberings about the pelican flock that stayed through last summer on Keystone Lake.

Then his mind turned to the individuals. "Dad, do great blue herons ever come in flocks?"

"No," I answered, purposefully forgetting the rookeries that are rumored to exist out here, purposefully equating that most impressive of creatures with a majestic sense of independence in his young mind. "I've never seen a *flock* of great blue herons anywhere."

"Especially in town," I could have added, "have I never seen a *flock* of great blue herons." But out jogging one day along a subdivision sidewalk, I spied a great blue heron, treetop height, heading into the northwest. Now *there* is one that flies its own course, I thought, so stopped for what seemed like such a long, long time just watching a great blue heron in town, tasting its individuality with my eyes, knowing the treat was mine. Down the block a man was doing something in his yard. I almost yelled at him to watch the heron, then didn't. Too busy to check the sky for heron, he didn't deserve the treat. Later that evening, Jenifer, our youngest daughter, came home from school. She must have thrown her things on the kitchen table and headed for the nearest basketball hoop; she was nowhere in sight. But scrawled across the front of her math book were the words: *Dad, I saw a great blue heron on the way to school today.*

So another has learned to look for that touch upon the air! Raising baby starlings taught me to look for that touch in birds; should raising baby people teach one to look for that touch in humans, that unique set of movements through which each of them becomes free? And can you see those movements from the earliest times in an individual life, can you detect the outcome? Or more significantly, can you force the outcome through application of opinion? No, to these last questions, no. You cannot produce a doctor, lawyer, professional athlete, through wishes any more than you can produce a pelican from a baby starling through misidentification. But you can have faith that those movements through a young time, so analogous to starling's first scramble up to the lip of a rotten cottonwood hole, will eventually reveal that touch upon the air required to get out of the hole. And you can, as human, influence the options, the avenues, the patterns those young wings fly, but only through the provision of many such options. A mother bird can never do that. There is only one avenue to the air, to the port of fledging. I learned all these last deep philosophical things by again standing out on a gravel road in the heat of a dusty day, watching for one more time that most familiar of sights: the cliff swallow colonies along the Sutherland Canal.

You all know about the Sutherland Canal; I wrote about it in another book. And you all know about the swallows around the spillway where that canal gets its water; you read their story in that same book. But what you never read was the story of those swallows two or three miles down the canal. The birds at the spillway live in a country club compared to those others. Down the canal, life begins to squeeze in on baby cliff swallows, especially when the spillway opens wide to drain the big lake. This happens every summer, of course. The dates are about the same every year. Some people use a calendar to tell those dates; I use the cliff swallow colonies a couple of miles down the canal. The spillway is opened about fledging time. When that happens the water level in the canal rises. When that level rises, it comes to within a foot of the bottom of concrete walkways a couple of miles down the canal. Underneath those walkways are the swallow colonies. Upon fledging, a baby swallow must make it out into the air, through that foot-wide gap between cold concrete and colder rushing water, *the first time.* Failure means death.

I've often wondered how it must feel to be a cliff swallow nestling. There is of course, the comfort of the flask-shaped mud nest with all its fleas and bugs. But then there is also the view of the world through a mud tunnel. Now tunnel vision is a common ailment, but it must be most common among cliff swallows. Furthermore, a nestling must make a transition so few others with tunnel vision can make, in order to fledge. That nestling must see the light and move toward it, moving in the process from a tunnel-vision view of the world to the lip where a panorama of choices awaits, then must make that first leap into those choices. That nestling/fledgling's equipment for surviving the leap into choices is a set of untried wings.

No wonder so many perish and no wonder the survivors are strong. I am most amazed, however, that in the case of another species, a species that I help fledge by the thousands, a species with infinitely more intelligence than the cliff swallow, with thousands of choices beyond that of the swallow, with a consciousness making all choices interesting, that we somehow seem bent on imposing that tunnel of mud, bent on limiting opportunities to try wings before making the fateful leap. That species, of course, is the *College student*.

Every week I go three times into an intellectual nest and see those babies crowded elbow to elbow into a mixture of expectations that simply cannot all be fed. For the record, the place is called Henzlik Auditorium. I am quite sure Henzlik made contributions. The whole building is named after him. Aside from the intellectual nest, there are dozens of offices and classrooms. These offices are filled with great teachers who teach how to teach: the Department of Secondary Education occupies Henzlik Hall. The *business* of Henzlik Hall is the ritualized, *thus strictly human,* passage of the accumulated history of the species. Henzlik would not have his name on that building had he not been a great one. But the auditorium, my intellectual nest, was not ever built for the ritualized passage of mankind's cumulative intellectual history. Henzlik never sat a hundred feet from a person who was trying to tell him something complicated, teach him a new language. Henzlik knows not what passes for teaching in the auditorium that bears his name. Nor does he know *why* we now use that auditorium. If he knew either, he would stagger in his grave.

Periodically I go to watch another parent feed the brood. I stand in the dark behind the last row, on the far edge of the intellectual nest with a hundred others, and try to get the strength to test some wings of the mind. I invariably come away convinced I know why a certain fraction of those nestlings is doomed from the moment they enter the nest, doomed to hit the cold concrete, doomed to hit colder water, unable to find the light at the end of the tunnel. We had a chance a few years ago to build a nest, a lecture hall where all could see and all could hear and all could receive the bountiful insects of wisdom and none would be crowded out of their opportunities. We chose instead to build a place to play basketball. It now takes three times as much money every year just to maintain that basketball arena as it would have taken in the first place to build a proper nest for over a thousand nestlings a day. Instead, I now try to feed the hoard in Henzlik Auditorium with poor Henzlik himself writhing in his grave over what politics hath wrought of the profession that admired him so. And out on the interstate, headed west at the end of spring, I ponder the false economy that dooms a fraction to failure and frustration. And at about Grand Island, with *Erma's Desire* approving on my left, I begin to get angry at false economies. And by the time I reach the Platte, the dusty road by the canal, to see those nestlings dead on the sand, see them swept away in cold water, my anger explodes at the false economies of politicians dabbling in education. I usually then pick up a flat rock and see how many times it will skip across the water. Passers-by think my mind is at ease.

Speaking of mud tunnels, now listen to this! One time a forty-year-old woman came to our colony. She was attractive and successful in her chosen line of work. She came by invitation and was paid a bunch of money to give a speech. There was evidently something about her line of work, maybe about the things she did in her spare time, that made a lot of our baby swallows want to hear what she had to say. So they paid her. She was paid with funds collected from all for the purpose of bringing out into the Great Plains attractive and successful people to talk about the things they'd learned during their lives. Those kinds of people have a way of widening the view from the mud nest, of helping to see beyond the Sutherland Canals of everywhere. The old swallows didn't like what this woman said to the babies. A few old swallows decided the tunnel should be kept intact, that vision from the nest should be narrow. These old swallows decided forty-year-old women like this one were dangerous, so the babies could no longer spend their collective monies to bring dangerous women into the colony. Babies squawked, to no avail. The tunnel remained intact for months. Months turned into years. No babies died because of the forty-year-old woman's visit to the colony. No babies were denied jobs, no babies made bad grades, no babies turned to hard drugs or burned anything or shouted *"Don't* GO BIG RED" because of the forty-year-old woman. But the experience of having the tunnel "mended" revealed more about the ways of the world than the babies bargained for. What the babies saw from their nest was the way old swallows reacted to *one* attractive and successful forty-year-old woman giving a speech. So what's new about all this? Nothing. But now when I go back to Keith County and walk the dusty road by the canal, see nestlings dead on the sand, swept down the cold water, I wonder if they would still be alive had their view of the world not been so narrow before they had to make that first leap. Then I skip another rock across the canal. But of course you're still wondering who was that forty-year-old woman? Her name was Jane Fonda.

And then there was still another time. Ann and I were doing what we so often did that summer—seine the South Platte at Ogallala—when I felt some eyes upon my back. There was a special channel along the north side that year, meandering along, making a sandbar after the floods, cutting up under the sitting spot. Twice a week that year we worked that channel for killifish. The swallows were mainly down at the other end of the bridge. It's been a long, long time since I was first hypnotized by those swallows in that particular place; so long, in fact, that they are no longer a distraction from seining—unless, of course, one is staring at you so hard you can feel it in your back. It is very hard to concentrate on a rivulet when you are being watched, analyzed, stared at, when an electric volt of curiosity sparks up and down the back of your legs. I tense as I write this; the feeling returns. The same tightness of leg muscles is with me now as it was that day. *Be careful! Be alert! Stop!* The gray mud of a postflood filthy river was drying to a plaster on the right bank. The branches of a fallen cottonwood, stripped leafless, tangled, seine-ripping, lay half buried. There among the branches were eyes that watched with such intensity. A cliff swallow fledgling was staring at us from that tangle of brush. Maybe it was only my imagination, but the intensity of that stare went so far beyond the curiosity or intensity of any human stare. A man and a girl seining killifish! Into what world had that fledgling been allowed!?

". . . into what world had that fledgling been allowed . . ."

There is something you need to know about perfectly fledged birds. They are clearly among the most beautiful of creatures. They have a special plumage, one that gives the impression of being untouched. It's the bloom on a perfect Cabernet grape, it's the ephemeral light of a pre-dawn March, it's the finish on some ancient painting. That special plumage is all the more special because of its owner's innocence. Both, you feel, will soon be gone. The innocence of a fledgling cliff swallow does not last long, one way or the other. We edged closer; the bird in the brush never moved. Its eyes were filled with the fire of curiosity; its head turned slightly to follow us. Its body was motionless; only the eyes, the eyes that met and fixed our stare, were alive. We backed off, almost afraid, quiet. That smallest of bundles of elemental curiosity was so near death.

When you hang around cliff swallow colonies for any length of time, you can't help being caught up in them philosophically, biologically. The wondering, thinking, the romantic analysis of a cliff swallow colony is something everyone does when given a chance. The brutality of a swallow's life is also shared. in the mind, by all humans nearby. There are always nestlings on the sand. There are always broken eggs, yolks dried. And there are always fledglings who don't make it. And for every such fledgling, there is a person who will pick it up and try to nurse it back to life. It never works. A streak of mud on that special fledgling plumage, a wing in the water, a collision with the bridge abutment, an exhaustion suddenly become too much for a first time cliff swallow chasing bugs that won't be caught—none can be countered by a human. But still there are always people who will pick them up. Oh, they'll eat, all right, back home, but not really for long. Their temperature goes down; they tremble finally and die, dried, frail, weighing so much less than they seemed to alive. Ann and I backed off and seined another part of the river, downstream a hundred yards, far enough away not to disturb the fledgling in the bush. But close enough, we were also, for those eyes still to see life in another universe: a man and a girl seining killifish. Into what world had that fledging been allowed!

A couple of days later the bird was gone. I reconstructed its fate in the fiction chambers of my mind. It had licked the mud tunnel, made it out into the air. It had seen the humans seining killifish and been caught up so deeply in its own curiosity that it came to the snag inches above the South Platte River. It never thought once about its own special plumage, its own vulnerability, innocence. There was something in that river it wanted to see. There was something new, some whole way of doing things, it never knew existed. What opportunity there is for speculation in this state of mind! As a parent, that swallow brings its own fledglings to the spot, hoping to show them a man and girl seining. The *idea* of purposefully introducing youth to strange ideas becomes more important than the man and girl. The colony does not agree; strange experiences are an irritating but normal part of youth. They should be forgotten, not encouraged. There is *colony* work to be done, tunnels to be repaired, when one gets a year older. Then one day I walk to Henzlik Auditorium in the late spring and there is that swallow chasing insects of wisdom around the front door.

This chapter needs to end with the story of how Jim Krueger decided to become a college professor. Oh, he's a long long way from being a college professor now, but it's the decision that's the important thing. This chapter also needs to end on a little more optimistic note, perhaps with a successful fledging, some example of good intentions turned into good, maybe the kind of chance happening that makes some person want to become a college professor. Jim may have had many reasons for making such a decision. But he would never admit the *real* reason; he may not even realize the real reason himself. I know what that reason is, however, for I watched it happen. Jim Krueger is the reason Harry Heron is alive today, *if* he's still alive; and Harry is the reason Jim wants to be a college professor. Make a difference in somebody's life, then you're hooked. How obnoxious can I get on this subject? Try this: Those who can help fledglings make it out into the world to function as free individuals, *teach;* those who can't, go into business.

Harry Heron came into camp in a box. The first thing he tried to do was get cold and die. The second thing he tried to do was stab Jim's left eye out. He fell a little short of success in both cases. One leg looked like it had been broken and Jim thought he'd hurt it the night of the storm. Neither of us was a vet but we finally decided it was a healed break. Harry's forehead was bald, too, which was kind of strange for such a youngster. But then Harry didn't *look* much like a youngster. I guess you could say he was one of those types that looks ancient regardless of how old they are. Maybe it was the fact that he was so beat up that made him look old and worn. Harry was also quite crotchety. He smelled pretty bad, too, and it got worse. In fact, the more I try to remember what Harry was like, the grubbier he becomes. I will say this for him: he finally did become paper-trained. He accomplished that when people moved him over to where the papers were and kept him there. All things considered, he was the filthiest, dumbest, ugliest, most bedraggled, hopeless case to ever arrive at the Cedar Point Biological Station, and my friends, *that's* saying something. Well, with all those things going for him, you couldn't help falling head over heels for Harry Heron.

It was probably the fish that finally turned Harry around. The first time he actually ate a minnow was the time we all decided he might not die after all. He was force fed for a while before somebody thought up the brilliant idea of giving him a fish. I mean, after all, what do Harry Herons eat out in the wild? You can bet your last dollar it's not catfood and raw eggs. But catfood and raw eggs he got anyway, in a dish sort of like in that fable about the fox and the stork. Of course he never ate any of it. So Jim would come in after a hard day's work and go force feed Harry and everyone would gather around and laugh and giggle and make all kinds of stupid comments. But you know way down deep in your soul that they were all wondering how they could have sunk to such social depths of associating with Harry Heron. That's when Harry went for Jim's left eye. Jim displayed his wound rather proudly, posing for pictures. But love has no bounds, so Jim went right back to forcing catfood and raw eggs down Harry's scraggly neck. Then someone got the bright idea of putting a couple of dead minnows on Harry's catfood. Maybe he would peck at the minnows, get some raw eggs and catfood in his mouth by mistake, find out how good it tasted, and by that convoluted set of events learn to eat by himself. Harry waited until everyone got disgusted and left before he quietly snipped those minnows off the catfood. After that it was a big game: people would give Harry a bunch of minnows then go off to watch through a window to see if he would eat. He did. And it wasn't but a couple of days later that Harry got to where he would eat with others around. And in a couple of more days he was eating out of your hand. Then the girls would spend a lot of time feeding Harry and the guys would spend a lot of time catching fish and you don't have to be much of a biologist to analyze *that* kind of social structure!

He came into camp in August after a storm. Jim was doing a

". . . the filthiest, dumbest, ugliest, most bedraggled hopeless case . . . you couldn't help falling head over heels for Harry . . ."

project on a great blue heron colony in the isolated tall cottonwood grove down the north side of the Keystone-Paxton road. It was one of those places that if you know anything about ticks you don't go into. It must have been an old farmstead, so isolated. Under the trees it was very still. The cottonwoods must have been sixty feet high. There were eight nests in that grove, all as high and spread out as they could physically be. The adult herons would leave when you were a mile away. So sometimes we went out there at night. But the adults would still leave; you could hear their guttural croaks fade away off in the darkness to the south. Then you could shine a light way up into the tops of those cottonwoods and there would be all of Harry's brothers and sisters and cousins standing riveted to limbs, straight, tall, so big, it seemed, so big up there in the dark. And you would look at those big baby herons and think what a job it must be to raise that kind of a bunch. Then you would not be surprised that great blue herons are not the most numerous of birds. You could stand there in the darkness with your flashlight out feeling those ticks crawling up your legs, a gentle rustle of cottonwood leaves a mile above up in the chapel roof, and imagine the storm that blew Harry out of one of those trees and into our loving arms. So after the storm Jim went out to see if his project was still alive, and lo and behold, what have we here but a *different* kind of project! There was Harry limping around and struggling through the bush. So Jim picked him up and brought him home.

Of course it's not too difficult to supply fish when you're out in Keith County. Everybody was seining for one reason or another. Even Ann's master's thesis research got recycled, thrown into the freezer and fed to Harry. There must be some of you who have picked up strays or other bums and blots-on-the-town. All manners of experience, literature, folklore, warn of the dangers of feeding strays. The altruism gets pretty tarnished sometimes, especially when the stray recovers faster than your wildest dreams, then gets demanding and unappreciative. Of course that's exactly what happened with Harry. Every fish that went down his stupid gullet seemed to make him more alert. Wasn't long before Harry seemed to assume that he had a *right* to be fed constantly. By the time mid-August rolled around and everyone started packing to go back to the city, Harry'd gotten to the point where he thought it was a *privilege* he was granting us, this *privilege* to feed him. And of course since it was a *privilege* to feed Harry, everybody outdid themselves, then the next thing you knew the privilege grew into a status symbol. All that sociobiology worked fine at camp. It started to break down right at the end when everybody realized there were only two choices: let Harry loose, which would mean sure death, or take him back to the city. So once again Jim bit the bullet and packed up Harry right along with all his other stuff.

We all waved good-bye to Jim, standing beside his little blue car packed to the gills out on the loading deck, Harry perched up on the back of the seat. Jim sort of looked reluctant to get into the car. We all assumed it was because he'd had such a good time at camp he didn't want to leave. In retrospect I think it was because he didn't relish a three-hundred-mile drive through that August heat with Harry for company. I sympathized a little bit with Jim. I remembered pulling into a filling station in Oklahoma with a redtailed hawk sitting up on the back seat, and buying regular for 19.9¢ a gallon. I remembered the hawk was rather tolerant but seemed anxious to get on with his own life. Harry, on the other hand, looked positively excited at what the city might hold in store! You could almost hear him calling, "Jim, take me to the city! Jim, take me to town!" Well, Jim took Harry to town, all right. Had to find a landlord that would take herons, said Jim. I mean Harry wasn't *really* a pet. Nor was he *actually* a kid. He was sort of halfway in between, with the worst instincts of each. A couple of weeks later Jim came into the lab.

"Like to borrow your seine," he said.

"Why don't you take him to the zoo?"

"They don't want him," said Jim. Evidently he'd made the rounds of places that might have some use for a great blue heron.

There is a name, Carol Odell, and a phone number written on my calendar, for use in just such emergencies. There is also a place called Chet Ager Nature Center out west of town. I gave Jim both phone numbers, but Harry ended up in Carol Odell's basement. Carol runs an organization called the *Wildlife Rescue Team.* Believe this, reader: Anyone who runs an organization called *Wildlife Rescue Team* possesses patience and tolerance that border on the inhuman. Such people are also idealists who defy description. Those kinds of people are vulnerable to herons. So Harry moved in with Carol, who let him try fishing after live fish in a tank. That, of course, made Harry so excited he started flying all around her basement, and *that,* of course, made Carol so excited she decided it was time for Harry to learn a bunch of more sophisticated things, like surviving out in the wild. Harry got rewarded for learning to fish and fly with a band on his leg. I can imagine that of all the banded herons in the world, only Harry is proud of his.

So Harry got taken out to the Nature Center in a box. He'd been brought into civilization in a box and was being carried out of civilization in a box. He stuck his head out, then stood up, then climbed up on the edge, then said goodbye forever to basements and cages and boxes and fish in tanks. We never heard much from him for a long time after that. Finally at Christmas I got a postcard from Harry. He was down in Florida with all his friends. He had a lot of comments about the crabs and seafood in general, the weather, about how he was having such a great time. His picture was on the front of the postcard. He looked pretty good. Even the bald spot had grown over.

Then right after New Year's Jim came into the lab. He'd shaved off his beard and looked downright civilized. We had a long talk about Harry, about how he was sure death until Jim rescued him, about how he was eating fish the size of your hand at the end, about how they banded him, about the postcard. Then we talked for a while about how Harry was finally doing what he wanted to do with his life, how his unique talents were being used the way he wanted, and about how Jim had played this crucial role in getting Harry from a down and out kid with no foreseeable future to a productive citizen of his own community. Harry had fledged, taken off on his own wings. Without Jim, Harry would now be a skeleton in the underbrush below those cottonwoods between Keystone and Paxton. I've been in this business too long without a break. It slipped right past me when I should have seen it coming a mile away.

"I've finally decided what I want to do in life," said Jim.

"What, Jim?" I had these visions of a life in which there was money, but not a lot, excitement in abundance, hope, a constantly changing and challenging life, with lots of freedom, a "job" in which you could almost dictate your own schedule, one you worked at because you *wanted* to and not because someone *told* you to, a career of ideas, not of tasks, in which your product was success. I couldn't imagine Jim in any other kind of life.

"I want to be a college professor," he said. "I don't know where or in what area, but I've decided I really like teaching."

I nodded and understood.

11

Local Color

ONCE every so often, when I'm working out in Keith County, I slip into town in the morning and drink a couple of beers at the Sip 'n' Sizzle. Those beers taste especially good at about ten A.M. on one of those hot, still days when you've gotten up very early, before daybreak, done a bunch of work of some kind, watched the sun rise over the McGinley Ranch, then done a bunch more work of some kind, then are ready for some thinking. So once every so often, when all those things come together in the right combination, I find some excuse to head into The Sip and down a couple of cool ones. There's always a bunch of locals sittin' around, some with a shot glass, some with coffee, just sittin' there in the sun shinin' through the Sip 'n' Sizzle window. Especially in July, they'll sit there and watch the harvest roll through town: trailers, combines, pickups. I wish every one of you who's eaten a loaf of bread could sit with me down at The Sip over a couple of beers on a hot, still July morning with the locals and watch the harvest move through town.

Blondena runs The Sip, now that Bill's gone to Montana to work on the oil rigs. She's been back behind the bar for four or five years now. She'll stand back there and watch the harvest roll through town, too, watch the characters roll in with the harvest, watch the migrants headin' on down to the coin-operated laundry. She's a pretty alert lady, Blondena, keeps track of things, remembers things. I've tried to trace all this history back for you, now that I've actually decided to tell someone, and have about decided that it all started on just such a morning sittin' around with all the locals talkin' about the harvest when Blondena looked out the window and said, sort of wistfully, sort of with that resignation, sort of with that strength she can get in her voice:

"Dinkle's back in town. Must've come in with the harvest."

All I remember about that particular morning is that all the locals got up and left very soon after that. Like a fool I stayed to finish my beer. Like a fool I was still sittin' there, sippin' that cool one, when the animal I now know as Lonnie Paul Dinkle walked through the door.

"Hullo, Blondena. Gimme the usual!" He had this pack of cigarettes he was banging as hard as he could against his left fist; I could see right away he also had this great gap between a couple of his front teeth. That's the kind of thing you notice when a guy smiles.

"What's 'the usual,' Dinkle? Last time you's in here I poured you ever' drink in the house!" Blondena has a way of getting her licks in early, sort of before things get out of control.

"Jack Daniels!" He watched her turn back among the bottles and she watched him out of the corner of her eye. "Hey, and a beer on the side!" She already had the beer half drawn when he said it. "You and all the other good people of North Platte will be pleased to know, Miz Blondena, that I've decided to settle right here in this be-yoo-ti-ful town on the prairies!"

"I thought you'd just come in off the harvest, Dinkle, how long you been here?"

"Just come in off the harvest. Come in off the harvest last year, year before that. Said to myself this time, 'Dinkle,' I said, 'you come in off the harvest too many times. One o' these days you'll settle down right here, so why not *now.*' That's what I said,

Blondena, said it, did it, and glad of it! I now got a *home,* Blondena, a *home,* pal! (He popped me on the shoulder hard.) Find myself some local girl, marry her, settle down right here, live the good life in North Platte, New-brass-key!"

"This ain't North Platte, Dinkle," said Blondena, sort of resigned-like, "this here's Ogallala."

"Can't matter a *whole* lot, can it, Blondena, can it?" He cast a squinty eye over the top of his shot glass.

"It do if you're in one town and that local girl you're plannin' to settle down with is in the other! Never find 'er, now would you, Dinkle?"

Then like some idiot I had to add my two cents worth. "You come in here convinced you're in North Platte, but you're really in Ogallala, Dinkle, means you're in North Platte in your *mind.* If you and that girl aren't in the same town in your *minds,* you'll never find her! What you need to do is go out there right now and find some girl *thinks* she's in North Platte!" You've probably realized from reading the rest of this book that I can't keep out of philosophical things. It'd help once in a while, however, if I could keep my mouth shut.

"You sound like an ed-yoo-cated man," said Lonnie Paul Dinkle, "and I want you to know you're talkin' to another! Good philosophy you got there: find yourself a girl's in the same town as you are in her *mind!* Reminds me of a teacher I once had in college."

"You never been to college, Dinkle," said Blondena. "This here's Dr. Janovy. He wrote that book about this place. Now *he's* a guy who's been to college!"

"I been to college," said Dinkle. "Oh, my God, have *I* been to *college!* Might go back one of these days, get myself a *Pee-Haitch-Dee* some day so's I can come into a bar in North Platte and have Blondena tell other folks I'm *Dr.* Dinkle. I tell you, Dr. Whateveryournameis, I got dreams, I mean *dreams.* I got *ambitions!"*

"I'm not so sure gettin' any *Pee-Haitch-Dee* is much of a sign of ambitions these days," I ventured. Lots of folks think it's only a way to security, state supported tenure, guaranteed dole for life from the taxpayer. Hell, I've seen these kids come along nowdays. Mostly a *Pee-Haitch-Dee* is only a ticket to a post-doc, which is then only a ticket to a chance to see if you're good enough to stay at some place so strapped for money it's tryin' all the time to get rid of you. Dinkle didn't exactly strike me as the type.

"I tell you, pal," said Lonnie Paul Dinkle, obviously oblivious to my second philosophical venture, "I seen it all. I mean *all.* Right, Blondena? *ALL.* I've been all up through the heart of this country watchin' folks make wheat, seen Calgary, seen Oklahoma City, met everyone, you name 'em, I've met 'em. Dumb kids, smart kids, *their* folks, ranchers, big businessmen, I mean *BIG* businessmen, own a dozen combines, women, I seen women. Boy! Have I seen the women. I seen women drivin' trucks, drivin' combines, cookin' out in the fields, I seen 'em ugly and I seem 'em be-yoo-ti-ful, I seen 'em on foot and I seen 'em ride a horse like you'd never imagine. I seen it *all,* pal, *all!*" He sucked on his beer hard, sort of like a combination period and exclamation point.

"You might of seen all those women, Dinkle," said Blondena, "but I'll give you ten to one none of 'em ever looked twice at *you!*"

"I been with the circus last winter," said Lonnie Paul. "Circus winters down in a little town in Oklahoma. Got back from Canada last year, couldn't find no work in O-K-C, so went on down to where this little circus was stayin'. See it all in a circus. You think you seen it all on the harvest, harvest ain't *nothin'* compared to the circus. Took my harvest money, bought a trailer and a pickup from some ole boy, lived in that trailer all the time with the circus, all winter down there in Oklahoma livin' in that trailer. I tell you, Blondena, that's the first *real* home I ever had."

I swear Lonnie Paul Dinkle's eyes all watered up. He blinked a couple of times then tried to wipe them so's we wouldn't see. We caught it anyway. Then he heaved himself down off that stool and went on back to the restroom. Lonnie Paul's jeans were awful baggy in the rear. Worn around the cuffs, too. Lonnie Paul's jeans were very worn right around the cuffs, sort of like they'd been draggin' on the ground a lot. And his shirt was out in back. In all the years I've known him since that day, I've never known his shirt to be in for very long. And I've never known him to wear any belt other than the one he had on that day; had the name "Maisy" carved into the back of it. And I've never known him to wear any other buckle than the one he had on that first day. "Keep on truckin'," read that buckle. I think that buckle probably says more about Lonnie Paul Dinkle than anyone else could. I also think he knew it; that's why he wore it. I had real trouble imagining Lonnie Paul Dinkle with a Ph.D.

"Hey, Dinkle," says Blondena when he returns, "who's 'Maisy'?"

"Maisy," he said, pushing his two glasses over toward Blondena, who was of course reachin' for 'em anyway, "was my first and only true girl."

"She must have been somethin's awful special," said Blondena. You don't just have *any*body's name carved in the back of your belt.

"She was the librarian down in this little town where the circus stayed. She worked with me day and night, down there in the library, helpin' me. I knew exactly what I wanted to do, been thinkin' about it since I was a kid, but she was the one, yep, Maisy was the one, knew all the poetry that would go with the animals. She was just like my mom," said Dinkle, "just like my mom."

"Was she as old as your mom?" asked Blondena.

"Almost," said Lonnie Paul, "but not quite. But she sure knew all the same poetry. See, I been thinkin' ever since I was a kid, been dreamin', about real poems and the stuff to go with them. Circus give me my only chance. Harvest give me the money, but the circus give me the chance. There's *no* place in this world you can do the things you can do in a *circus*."

I almost said something about being able to do as many things in a university as you can in a circus, but didn't. Didn't want to destroy whatever respect Lonnie Paul might have had for a Ph.D.

"Tell us about the circus, Lonnie Paul."

"Yeah," said Blondena, pushin' more glasses down on that wash brush in the sink, "tell us about the circus." She winked over at me.

"We took nursery rhymes," started Lonnie Paul Dinkle, "and put the animals with them. Done it all out in the center ring, lights out, big spot down on me dressed up as this professor, doin' them poetry readings, them animals goin' through their stuff. People loved it. I thought a lot last winter about bein' a *real* professor."

I kept awful quiet.

Then Dinkle goes into this act, reciting all this poetry, long nursery rhymes, all the verses, all the tragedy, love, success, frustration, hate, violence, of all the world you remembered as a little kid. And as the afternoon wore on you could just imagine all those animals he had acting out all this stuff and it turned grand and glorious there in the dim light of the Sip 'n' Sizzle that hot afternoon in the middle of some long ago summer. And the sun worked its way across the sky and the shadows worked their ways across the rows of bottles, and it got oh, so maudlin, and once in a while Blondena'd nod, and once in a while Lonnie Paul'd hit one my own mother used to tell me. Then of course I'd nod a little. But on and on he went, people walkin' by, puttin' their hands on the door, shadin' their eyes so they could see in, lookin' through to see who's in there, then wanderin' off across the street to the "Hoof 'n' Horn," havin' seen Lonnie Paul Dinkle at the bar, knowin' he'd come in with the harvest. Finally about four-thirty Lonnie Paul says:

"Got to go, Blondena. Laundry's probably ready."

"You put it in before you come down here, I'll bet you a beer it's ready!" Blondena sure is a practical woman.

"See ya, pal."

"See ya, Lonnie Paul, thanks for the memories. You know you just went through the whole history of mankind with all that poetry, all the joy, sadness, excitement, bravery, duplicity, all of it! Don't know about Blondena, but I loved it! Thanks!"

"Yeah, thanks, Dinkle," says Blondena.

He waved his hand. As he left, a couple of cowboys came in the back door. Blondena and I looked at one another and smiled.

"Is he gone for good, Blondena? Is this some dream come in with the harvest and goin' out with the mornin' fog over Big Mac?"

"Nope," said Blondena, "he'll be back."

"You're *sure* he'll be back?"

"Sure I'm sure. Lonnie Dinkle's just gettin' started!" Back in the poolroom someone broke and someone laughed and someone cursed and I heard a couple of balls go in and roll down in the rack, that rack from which only a quarter will get them back. And that was the morning and the evening of the first day.

The morning of the second day started as only those mornings in Keith County can start. You could tell it was going to be still and hot. Light came in clear over the ranch to the east of the station and you could hear the falls at the diversion dam half a mile away. Western kingbirds snipped and slashed the predawn air above willows down by the parking lot. You could hear some very gentle walking down through the brush and before long you could see a fly line laid out gently over the canal. Some student, I thought, come down to the canal for trout before breakfast. Down along the canal to the east, underneath the bridge, silhouetted against the fog, another was tossing a Meps spinner into the deep water right where the canal enters the lake. Off to the west you could see some campers creeping along the top of the dam, headed north. Some had boats trailing behind. There was a gray layer of midges two inches thick back on the patio; must've been a big emergence last night, I thought, but only for the most fleeting of instants. You guessed it: I had to find some excuse to get back to The Sip that morning. And you also guessed the rest of it: I should have stayed at camp, gone fishing, taken the kids canoeing, *any*thing but go into town. Needed to go in and pick up some mail, I said to some folks; expecting an important envelope from New York, I said to some others; got to check on our food order at Sixth Street Market, said to some others. Big joke. Fifteen minutes after nine I was sitting there on the stool that still bore the impression of my rear end from yesterday afternoon and sixteen minutes after nine Blondena set that Miller Lite in front of me. I ignored the cowboys in the back room; I had this feeling they were still there from the night before.

"Hey, wake up Dinkle," yelled Blondena, "Dr. Janovy's here."

Wake up Dinkle? I thought.

"Shoulda stuck around last night," said Blondena. "Dinkle came right back in about fifteen minutes after you left."

"Guess I figured we both had enough for one afternoon," Thinkin' of the poetry, of course.

"Got his trailer stolen from down by the laundry," said Blondena. "Guess they took it all. Poor guy's got nothin' but the clothes on his back."

I was beginning to get the picture.

"Stayed here, he did. Guess I'm just a softie." She pushed some dirty glasses from the night before down on that washer brush in the sink.

"Did you call the police?"

"Dinkle don't like to be around the sheriff very much."

"I didn't realize you had a place to stay in here."

"I don't." She sort of looked up over the bar and into the back room.

I turned around and looked past the jukebox, through that arched doorless portal into the wages of billiard sin. It took me a while to realize there really was a pair of boots lying up on the pool table. I guess I finally understood that when one of the cowboys banked a shot off Lonnie Paul Dinkle's boot toe. Cue ball rolled very gently over and tapped the three into a side pocket. I sort of sauntered back into that back room. Cowboys hardly noticed, they were so caught up in the heat of competition. Dinkle was snoring loudly, strung out up on one side of the pool table, muddy boots up on the felt, arm sort of droopin' off the end at a crazy angle.

"We calls it *three-pocket!*" says one of the cowboys, spitting some snuff down in an empty beer can, chalkin' his cue. "New game."

"Yeah," said the other, takin' a big dip of Skoal then right after a big drink of Bud from a warm can, "it ain't only new, it's the only one we can play with ole Dinkle up on the table!"

"Nice shot you made off his boot," I ventured.

"That ain't nothin'," said the cowboy. "You ought to see the one off his belly."

To prove it he banked a shot off Dinkle's belly, just missing the "Keep on Truckin'" belt buckle, so the cue ball just sort of died out in the middle of that green felt with all the beer stains. About this time Lonnie Paul begins to stir.

"Sweet dreams, Dinkle?"

"Kept dreamin' I was out in some hail storm in the middle of harvest and couldn't get away; hail as big as golf balls, hittin' me all over, feet, body, all over," said Lonnie Paul Dinkle.

"One off his forehead's easier to make than the one off his belly," said the taller of the two cowboys, the one with "Dave" carved into the back of his belt.

"Shame about your trailer."

"Gone, ain't it." He started lookin' around for his cigarettes, pattin' all his pockets. "Need a little breakfast." So he reaches over and takes a big drink out of a beer can. Wasn't but a second or two after that I went back up to the bar.

"Coffee, Blondena." Dinkle was right behind me.

"You live on the road a while," he said, "like me, sleep all over, talk to the common people, you get a lot of ideas, learn to do a lot of things, right, pal?"

"Right, Dinkle. What are you going to do about your trailer?"

"Didn't call the sheriff, did you?"

"No. Where are you going to stay?"

"Where are *you* stayin'?" asked Lonnie Paul Dinkle. "College professor don't *live* out in these parts, right? *You* got some place you're stayin'."

"I live at the camp." Then right away to change the subject I started talkin' to Blondena about the windmill. They're puttin' this gigantic windmill up on the bluffs. An energy experiment. Blades bigger than you can imagine. One of three in the county. I think they picked the right place, up on that bluff. Of course you also have to realize that Lonnie Paul is not truly in control of his own mind. Start talkin' about something like that windmill, next thing you know he's right in the middle of the conversation. I wasn't at all interested in pursuing the bit about a place to sleep.

"That windmill out near where you're stayin'?" asked Dinkle.

"You wouldn't believe the size of those blades. Why, they've got a place big as a feedlot fenced off just to keep people away from those blades. And fast, you wouldn't believe how fast those things turn. And electricity, you wouldn't believe the electricity those things generate, when the wind blows, that is. Why that electricity just comes sparkin' over the wires, heatin' up water all over the county, runnin' air conditioners. I'll tell you this, Dinkle, we're on the right track with those windmills."

"Must be somewhere near where you're stayin', pal?"

"Took 'em a year just to build the first one. Now they've got that one up, everyone wants one."

"You're so familiar with those windmills," said Lonnie Paul, "they must be right near where *you're stayin'*, right?"

"Build five or six of those things, could probably put electricity *back* into the lines, wouldn't be buyin' it, we'd be *sellin'* it," I said.

"Then you'd be the same as the yoo-till-i-tee companies," said Dinkle, "with your own little yoo-till-i-tee right out there where you's *stayin'*, right, pal?"

"Probably."

"By the way, exactly where is it you's stayin'?"

"Dr. Janovy lives out at the camp," said Blondena, bless her soul. "Ever'body knows where that is, out at the old Girl Scout camp. Dr. Janovy's *in charge* of that camp." Generally I've found it's a mistake to let anyone know you're *in charge* of anything. It never occurred to me maybe I'd better warn Blondena. Go around warnin' people not to tell anyone you're *in charge* of something, that's the same as tellin' em you're in charge of something, right? Of course.

"IN CHARGE?" screams Dinkle. "You're *in charge* of a *camp?"*

"The university runs a summer program out north of town."

"And you're *stayin'* out there at a *camp,* right, pal?"

"Right. Along with my family, all the students who've paid."

"Paid," said Dinkle, *"paid, paid, paid.* Well, get a place in camp *ready,* Doc, 'cause old Lonnie Paul Dinkle is comin' to camp with you and all them college kids!"

"You got money?"

"Work it off. Camp's always got need of help. Work off my room and board."

"Lonnie Paul, we got a full-time maintenance person. Not only that, he's the best there ever was."

"He needs help."

Now I've not been *in charge* of all that much in my life, and even what I have been *in charge* of, I've not been *in charge* of it all that long. So you can understand why I didn't pick it up right away that Lonnie Paul Dinkle was not the kind of person who responds to subtle hints that he's not needed. Not only that, college professors are notoriously bad managers. Maybe that's why higher education is in such bad shape—most of the administrators are former college professors. Besides that, I'm a real softie, pushover, easy mark, ultimate touch, the guy who always says "yes" and always gets told "no." I wonder if all college professors are like that. I wonder if all those college professors who've become college presidents have hired their own Lonnie Paul Dinkles. No wonder higher education is in such bad shape. Maybe there are thousands of Lonnie Paul Dinkles disguised and running around high offices in all state universities. I can't judge any of those possibilities. In addition to being an easy mark, I always seem to surround myself with the kind of people who think there are not enough Lonnie Paul Dinkles running around in the upper administration! There are even some days I think they're right! Anyway, I made a real effort to leave smoothly. Back out in the hundred-plus sunshine of midmorning Ogallala I unlocked my car door and was struck by a furnace blast smelling of stale cigar. That's usually what happens in the summer in my car. I rolled the windows down as fast as I could—an incredible mistake. No sooner had I started the engine than Lonnie

Paul Dinkle opened the door and sat down right beside me. My personnel problems had just begun and Dinkle was the stone that rippled the waters. Even after all these years, he's still doing that. It just makes me cringe to hear those stories of how he's doing at Notre Dame, Texas A & M, Tulane, Wake Forest, all the places he seems to end up.

"Are you planning to put him to work helping *me?*" Phil would ask later on that day. Phil is the best maintenance man any camp ever had. Something in the tone of his voice made me answer no.

"Is he planning to pay room and board?" Monica would ask later on, receipt book in her hand. Monica is the best secretary/accountant/executive assistant any camp ever had. Something in the tone of her voice made me answer yes. Lonnie Paul's still not paid his room and board, in cash, that is.

"Are you planning to put him to work in the *kitchen,*" asked Cynthia, big serving spoon held sort of like a machete. Cynthia is the best cook/kitchen supervisor any camp ever had. Something about the tone of her voice made me say no. But those things all happened later. Wasn't fifteen minutes after I'd left The Sip that morning with Lonnie Paul talkin' my ear off, that we pulled up in front of the white gate to camp. Of course I stayed in the car. If Dinkle's been on the harvest he knows his manners, I thought, and at least I was right about that. He hopped out right away to open the gate. He fiddled and fiddled with the latch and chain, but pretty soon he came back with this big ole grin, sort of foldin' his pocket knife up, pushin' the gate open, closin' it behind. Wasn't until then I remembered that Dr. Gainsforth had been lockin' the gate.

"Tricky," said Dinkle, back in the car. "Sometimes them laminated locks takes a key." I said nothing. Then I noticed all these vehicles parked up on the hill by the big windmill. *Central Nebraska Power and Irrigation District* said one on the side of its door, big four-wheel-drive pickup. *Midwest Electric* said another, bigger four-wheel-drive pickup. You could see up the hill a bunch of guys were standin' around spittin'. Dr. Gainsforth's Jeep was off to the side. About that time I realized those blades weren't turning. The wind was blowin' a gale. Of course, as camp administrator my first concern was about electricity. I'd have scooted on up there in a hurry except that Dinkle'd left my door open and was halfway up the hill. I turned off the engine but left Waylon on the tape player and watched Lonnie Paul Dinkle work his way up the hill through the brush toward that crowd.

"Who's that guy, Dr. J?" The vans had pulled up behind, in from some field trip. A student was at the car window. I was blocking traffic.

"That's Mr. Lonnie Paul Dinkle," I said.

"Reminds me of my English teacher," said the student. "What's wrong with the windmill?"

"Lord knows."

"That guy Dinkle think he can fix it?"

Somehow that line of conversation didn't seem to be leading in the right direction. Dinkle had disappeared into the crowd way up on the hill. I didn't want to say anything about the lock to a student; they get pretty touchy about stuff like that. I shrugged. About that time those big blades started moving, very slowly, then pickin' up speed. I could see some people gettin' into their pickups and one of 'em was shakin' hands with Lonnie Paul, way up there on the hill. I reached over in the back seat for my binocs and checked it out. Sure enough, one of those power district guys was shakin' Lonnie Paul's hand. Here came Gainsforth bouncing out of the underbrush in his Jeep and pulled up beside. This was all gettin' to be a real social occasion.

"Who's that guy came up the hill?"

"That's Dr. Lonnie Paul Dinkle," I said.

"Visiting professor?"

"Amazing guy," I replied without really replying—sort of hinting, I suppose, that Dr. Gainsforth had identified him correctly. "Dinkle is one of those guys who can do about anything. He'll win a Nobel Prize one of these days!"

"I believe it," said Dr. Gainsforth. "Anybody can fix one of those things with his pocket knife is Nobel Prize material!" He gave me that little smile and the Jeep bounced on down the road. And that was the morning of the second day. I don't want to tell you about the evening of the second day.

Wasn't too many days after Lonnie Paul comes to work at camp that we went back in to pick up his pickup. He'd left it at the laundry. He'd stayed at camp for a while—sworn off beer, he said, never goin' back to The Sip. Ever'body knows that's a lie; ever'body goes back to The Sip. Lonnie Paul put in his days workin' but then after a while asked me if anyone's goin' in to town. Needed to pick up his pickup, he said, and sort of grinned that Lonnie Paul Dinkle grin. I told him I hoped it was still there, thinkin' of course of his trailer's fate. He assured me it would still be there. Sure enough it was. Black, beat up, 1948 Ford pickup, different size tires on each wheel, all the windows down, seat all bare springs. He stuck his pocket knife in the ignition and the thing kicked right off. I left. Obviously Lonnie Paul didn't need any help with the white gate at camp. Later in the day his truck showed up in the parking lot. Way it was parked I could tell he'd broken his personal vow about The Sip. You could tell that too from the way he showed up at dinner—in cutoffs. Now our camp gets to be a pretty casual place, especially in July, but Lonnie Paul Dinkle in cutoffs just offended everyone, especially the faculty wives. I asked him to leave, go cover up those legs. He wanted to talk about wrens instead. I remember the evening very well. It was then we heard the first of what we've come to call a "Dinkle" around here, now that Lonnie Paul's long gone. "Dinkles" are those things high-class scientists wish they'd thought of first.

"Been watchin' those wrens down by the cattails," mused Lonnie Paul. "There's lots of things a guy could do to answer all these questions I got burnin' inside."

"Like what kinds of questions, Lonnie Paul?"

"Like what are the things that actually determine whether a nest becomes a false nest or a real nest. Find that out you might figure out what makes any animal's home a fake home or a real home." He gave me that grin and look, "Way those things live out there, could just save a bunch of old nests. Wait till next year when those wrens have their places set, then you could add those old nests, could add old real nests in place of the false nests, could remove their false nests as they build them, substitute old ones but in a slightly different place, could put some fake eggs in the false nests, could take some stuff and build that little platform inside a false nest the same way they build it inside a real nest. I could think of a million things to do."

"Maybe not a one of them would ever tell you why a wren makes one nest real and a bunch of 'em false," I said. Can't just have *anybody* comin' in here thinkin' up good ideas high-class scientists wish they'd thought of first.

"Maybe not," said Lonnie Paul Dinkle, "but you'd sure as hell find out *somethin'*." He picked up his tar bucket and headed back for the garage roof where Phil'd sent him. It must have been a hundred degrees.

"Been watchin' those spider wasps down by the road to the Gainsforths," mused Lonnie Paul a couple of days later. "Most fascinating animals I ever seen. Why a guy could take that whole bunch into some schoolroom. I seen hundreds of things out here a guy could just pick up lock stock and barrel and take into some schoolroom. Never seen a bunch of wasps act like that; never seen a bunch of *any*thing act like that. Tell me why, Doc, why they never have nothin' as fascinating as those wasps in schoolrooms?" There is a pile of sand drilled with spider wasp burrows down the road to the Gainsforths. Those wasps go off and sting a spider, bring it back and lay their eggs on that paralyzed spider. Eggs hatch and wasp babies have a ready supply of food—the spider. There is this elaborate ritual those wasps go through at their burrows. I'd spent my time fascinated by those very insects. I couldn't answer him. Money, maybe money was the problem. I don't know today why "they never have nothin' as fascinating as those wasps in schoolrooms."

"You finished diggin' out the rain ditch, Lonnie Paul?"

"No, Doc." He picked up his shovel and headed up the road. We have these concrete drains along the service roads. If you don't keep them dug out, rain washes out the road every time. It must have been a hundred degrees that day, too. He eventually got to where the drains needed to be dug out, but he stopped a couple of times to look at something in the grass.

Wasn't too long after the wasp business that I saw him way up on the bluff just sittin' there lookin' off into the Sandhills to the north. Got close to dinner time, we started yellin' up at him to come on down for dinner. Lonnie Paul acted like he never heard a thing. We yelled a lot louder and rang the big dinner bell. He still didn't even change the way he was sittin'. Lonnie Paul Dinkle wasn't in the habit of missin' meals. I went pickin' my way up the bluff to see if he was dead or somethin' and you have to know those bluffs are not the easiest things to climb. I'm always puffin' when I get to the top.

"Dinner, Lonnie Paul," I puffed, then took a couple of deep breaths. "Some reason you been sittin' up here all afternoon and can't hear the dinner bell?"

"Just look out there, Doc," he nodded off toward the north into the Sandhills. You could see maybe fifteen miles. "Doesn't that just give you some feelin' inside you can't describe? Don't you just feel you *need* what you're seein' out there?"

Read a scientific article not too long ago in which some guy had analyzed human need for wilderness. This article said that humans evolved in the wilderness, therefore had a basic need for exposure to natural areas before they could exhibit natural forms of human behavior. By comparison it said that a gorilla in a zoo cage is a dangerous and unpredictable neurotic, but a gorilla in the wild is a shy, tender, sensitive, gentle, and intelligent soul. The gorilla was said to have evolved in the wilderness the same as the human. So this article implied that human behavior in the absence of wilderness was apt to be dangerous and unpredictably neurotic. The article thus explains my own behavior that evening with Lonnie Paul up on the bluff.

"Don't you just feel you need *what you're seein' out there?"*

I said to hell with dinner, too, and just sat up there and watched the Sandhills to the north until dark. You get a feeling inside that you can't describe when you do that. It's not easy gettin' down from that bluff in the dark. Lonnie Paul's voice came out of the dark brush behind me.

"Anybody workin' to save these prairies, Doc? Save 'em in their original condition?"

"You mean so guys like us can spend a few hours just lookin'?"

"Yeah."

"A few, I guess. They're successful once in a while."

"Shame," said Lonnie Paul Dinkle. "From what I felt today I'd say they need to be successful more than once in a while." He picked up his mop bucket and headed for the lab. Lonnie Paul did a pretty good job of cleaning up after ichthyology that summer. "You goin' into The Sip tonight, Doc?"

"Yeah."

"Talk to me tonight about these prairies?"

"Sure, Lonnie Paul, any time. I'll talk to you about these prairies any time."

He almost always showed up for meals. Lot of those kids stay up through the night working on some project, studying for exams, then sleep right on up to thirty seconds before class. It's kind of fun to sit there with my twenty-fifth cup of Cynthia's coffee and watch those all-nighters come hustlin' down the path at about one minute to eight. But Lonnie Paul was never one of those. He was always at breakfast and lunch and dinner. A lot of times people would leave their extra food for him. A lot of times he'd get up and make a couple of peanut butter sandwiches to go with that big tray of great stuff Cynthia always dished up. Then he'd drink four or five big glasses of milk. I read somewhere adults are not supposed to drink milk. That sure never phased him. Our bill to Beatrice Foods that summer was enormous. He could put away the ice cream, too. Milk deliveries came in the middle of the night, Dan drivin' that gigantic refrigerator truck down our winding trails and narrow cattle guards. Lonnie'd almost always be there to help Dan unload. Dan'd almost always give him an ice cream bar afterward. Once in a while we'd see Dan in The Sip late after his deliveries. A few times Lonnie Paul would buy Dan a beer, just sort of motioning to Blondena with his head to set a beer over in front of Dan. I guess Lonnie Paul still had some money left over from the harvest. The camp account never saw any of it.

But he was a pretty good worker. He was willing to do a lot of things you could never hire any student to do. He did those things cheerfully, then went off for a couple of hours to do something for Lonnie Paul Dinkle. I wish you knew how many hours I've spent in the last fifteen years trying to get college students to do just that: go off by themselves and study something they choose out of personal interest. It's a rare college student that will do that. All the ones that have ever done that have long since become very successful—journalists, doctors, scientists. Of all the things in which I invest time, I've come to feel over the last fifteen years that the most worthy is another human who will purposefully choose what he or she is interested in, then commit the time to study it, usually with no guidance. I think this act brings into play the most unique and valuable of human attributes—the original thought, the original point of view, the original interpretation—all innocent and naive, uncolored by some professor's prior opinion, or by some obscure paragraph in some book.

The tolerance and patience I've shown in past years for people like Lonnie Paul is the price I've paid to be associated with people who think their own thoughts and choose the things they themselves want to study on their own. That tolerance and patience is viewed as one of my eccentricities. My mysterious ability to attract people like Lonnie Paul is viewed as one of my administrative weaknesses. It's just as well. I've been often frustrated in my competitive aggressive drive to achieve a position that brings out my highest level of incompetency. Others seem regularly to get promoted beyond their abilities; why not JJJr? I think it's because of the Lonnie Paul Dinkles with which I've been patient. Higher ups don't trust me. I associate too freely with those who cannot be "guided" easily in their thinking and doing.

So much for the philosophy of careers. We lost Lonnie Paul at the end of the summer, all of which made my normal case of the Ogallala Blues especially bad. They were bad anyway; something about that particular summer, the combination of weather, company, things that happened, people freed of all that city stress and hassle, people giving all these magnificent performances, coming up with all these ideas, no one telling anyone they were wrong, or couldn't think some thought, all those things came together in a combination that made the Ogallala Blues especially bad when it was over. Then we had to lose Lonnie Paul. Of course anyone would know it would have to end. It's just that there's something in your mind that refuses to accept the reality that it all has to end, that the frost will come along and you'll go back to the city and those beaches and people, and the spectacular physical beauty of the Nebraska Sandhills will only be unreal memories and dreams, something you did so long ago. Then it's back to all the crap. Lines and deadlines. Meetings. A calendar that seems to add obligations by itself. A hundred extra faces in class. Piano lessons. Forms to fill out. So it's the Ogallala Blues. What makes it all worse is when you yourself are in the middle of all this crap and you know that one Lonnie Paul Dinkle has avoided it. You'd have thought my normal life was like some Biblical plague to Lonnie Paul. He just sort of smiled when I suggested that with all his talents, abilities, and unique ideas, he maybe ought to come back and try college again. He just smiled.

"No, Doc, not your kind of college, least ways not yet."

"You could make it, Lonnie Paul. You could even make it into graduate school in biology, eventually. I've seen worse in graduate school."

"Maybe some day, Doc," he still smiled. "But not your kind of college, at least not right away. Goin' to hang around here for a while. Get my stuff together. Maybe take some correspondence courses, get my grades up, you know, like those basketball players. Yeah," mused Lonnie Paul Dinkle, those suspicious things starting to turn around in his brain, "yeah, like those basketball players. Just like those basketball players. Maybe I can get eligible for the conference championship of life."

"I think you're already eligible."

"Got any recommendations, Doc?"

"Spharagemon State; right outside of G———, Oklahoma. Spharagemon State will send you some correspondence courses, all right."

"Sounds good to me," said Lonnie Paul Dinkle. "Yep, sounds *real* good to me. Gonna get my stuff together one of these days, Doc, then I'll be back to see ya."

"Don't threaten me, Dinkle," I joked.

He helped us pack up the library, helped us inventory the kitchen stuff, helped us inventory the upstairs storeroom, put away all the stuff from the patio, then went through all the discarded shoes out on the patio until he found a couple of pairs that fit him. He threw those in the back of his truck. The day we left he was standin' up on the loading dock just watchin' it all, leanin' on a big trash barrel. Then we left. One minute we were there, standin' by the car tellin' Lonnie Paul Dinkle good-bye, the next minute we were in the car, engine running and gravel popping up under the floorboards, headed back to the city. It's the first hundred yards of the hardest trip I take.

"You guys have a good year," said Lonnie Paul, standin' up on the dock.

"You get your stuff together, Lonnie Paul, win a Nobel Prize, y'hear?"

"Sure, Doc. Sure."

Then we drove off. Phil called me from camp in a couple of weeks, wanted to know if Dinkle was going to stay out there all winter. I told him as far as I was concerned Dinkle could stay forever. You could almost hear Phil shrug over the telephone. Not too many guys would even act like they were going to stay out at camp over the winter. Then Phil called me a couple of weeks after that; wanted to know what had happened to Lonnie Paul.

"His truck hasn't been here for three or four days," said Phil. I love talking to Phil long distance over the phone from Ogallala; there's all this cross talk on the lines, and you can pick up all this good Ogallala gossip from the long distance cross talk. "You know if he's planning to be back in here? I sort of need to close this place up."

"Don't worry about Dinkle," I told Phil. I didn't want to say anything about the uselessness of "closing the place up." Many moons have gone by since all this story took place.

So why am I telling you all this stuff about Dinkle now? I just came back from some scientific meetings and had to tell someone, that's why. I have this friend who's on the faculty down at Texas A & M—Mike Kemp, an immunologist, a very good friend. Every time we go to scientific meetings we get together for some good times and heavy philosophical conversation. Sometimes when he brings Kathy and I bring Karen then we'll go out to eat together, talk about the good times and heavy ideas. Mike's a real character. He once wrote a paper called "The isolation and purification of human soul," published in the *Journal of Irreproduceable Results,* giving all these biochemical methodologies for the purification of human soul. He kind of has that ever-skeptical Texas way of talking, all the mannerisms of a combination Dennis Weaver and Clint Eastwood. I got into Lincoln last night on the late United flight from Chicago, Mike Kemp's question burning in my brain like some kind of prairie fire.

"Ever hear of a guy called Lonnie Paul Dinkle?" asked Mike Kemp, my good friend, while we were sitting in the lobby of the Hilton in Chicago a couple of days ago.

"Why do you ask?" I was pretty damn wary.

"This guy showed up in my lab," said Kemp, "wants a job as a technician. Showed me his transcript; full of correspondence courses from some little place up in the Oklahoma panhandle. Said he knew you."

"I know Lonnie Paul Dinkle. You got the guts, Kemp, hire him."

"I already did," said Mike. "Anybody says he knows you, that's good enough for me."

"Tell you a little secret, ole buddy, one of these days Lonnie Paul Dinkle's going to win a Nobel Prize."

"Well," said Kemp, "it won't be for the stuff he's doin' now. I put him to work on the isolation and purification of the human soul."

So Lonnie Paul's made it as far as Texas A & M. I am resisting, truly resisting, the temptation to point out that he may be the ultimate Aggie joke. There are educators and there are educators, but there are none like Mike Kemp, the man who put Lonnie Paul Dinkle to work on isolation and purification of the human soul. Only one person in the whole world would perfectly match such a student with such a problem. I've never been to Texas A & M; maybe that's the only place in the world such a match could occur. So where is all this leading? I wish I could say to the end of my contact with Dinkle. He'll be back. Or maybe I'll leave Nebraska one of these days and run into Dinkle out there in the cold cruel world, out there with him running around equipped with all this training from Texas A & M, and Dinkle will get that squinty look on his face and say, "Hi, Doc, got time for a beer? Got lots to tell you."

And I'll say, "Sure, Lonnie Paul, I *always* got time for a beer with the man who finally isolated and purified human soul."

But Lonnie Paul will say, "It was all Kemp's idea."

Then I'll say, "Yeah, Lonnie, but you and I both know that not just everyone can come along and isolate and purify soul. It takes a special kind to do that."

Then he won't say anything, just get that look on his face, maybe staring off through some bar window out on the street of some town, just thinking. Then I won't say anything either, just sitting there like we did so many years ago back in The Sip, lookin' out that window, just thinkin'. But he'll be thinkin' about why he's never seemed to fit too well or stay too long in any place, and I'll be wonderin' why we never seem to attract many like Lonnie Paul Dinkle, or if we do, why we can't hold them. Then I'll probably conclude that even if you attract them, you can't ever hold them. They fledge, and fledge, and keep on fledging, forever and forever, and when it's all over they never seem to have much in the way of material possessions, they always seem to make people more comfortable when they're leaving, they're always talked about. But after a few years we begin to sort out the unique qualities of those kinds of people and forget the irritating stuff. Takes a bunch of patience, though, to be around them when they're in full swing. Maybe that's the patience required of a teacher. Maybe that's a patience so few of us have, and maybe the fact that so few of us have it is the reason the unbridled minds of Lonnie Paul Dinkles are such a rare thing.

Well, enough of this tale. I get all maudlin just writing about those kinds of misfits. Like to go on now to the subject of owls, a truly sad choice of a place to live, but in the end, an unforgettable lesson on the fragility of nature.

12

Owls

"BURROW owls," little John called them, instead of the proper "burrowing owls," and he'd raise up in the car seat straining against a safety belt just to see the babies. Then we'd get the tally: one through maybe five, with the adult up on a utility box, you know the orange kind that signify underground cables, stuck off in volunteer wheat along the bar ditch. That would all be going into town. The view was not so easy going into town and a person always had to decide at the last minute whether to YIELD to merging traffic or look way across the median for owls. We always chose the owls. Lucky for us, all of us that summer, I suppose, for I never knew a person who traveled that corner that didn't look for owls, that there never was much merging traffic to yield to in that part of the country. But of course most of the people I know, if given the choice between merging traffic and merging owls, will yield to owls. But then I didn't know everyone who traveled that road that summer. Some obviously didn't yield to owls.

The trip out of town was a good deal more exciting, and little John would begin straining against the seat belt a mile before the nest, for two miles north of town highway 61 takes a split-lane curve to the right and the owls had nested within a few feet of the asphalt. Maybe it was an old badger hole, who knows, but everyone I know who drives these roads looks for tell-tale holes like that then looks for a burrow owl on the mound in front. We weren't much different that summer, especially after we discovered the nest and even more especially after it became obvious it was an active burrow. Of course the real excitement started when the young began poking their heads up over the mound in front. That's when we'd get the count. Little John would count baby owls and I'd watch the oncoming traffic in the rear view mirror, calculating carefully just how much I could slow down and not scare the owls back into their hole, just how much I could slow down and not get hit from behind, and just how much emotion I could invest in an old badger hole full of little owls. I never got hit from behind and I rarely scared them back in. The emotional investment is a different story.

But then I wasn't the only one who miscalculated the emotional investment required to drive past, within a few feet, a burrow owl nest every day and every night for three months. There were a hundred of us and we all watched that nest. In the end we adopted them; no, not physically, but emotionally, personally, and in the end there were a hundred guardian angels watching that feisty brood. Early in the morning, in after mail and groceries, out in maybe the pelting blowing rain of the prairies, out in suffocating hundred-plus heat at ten in the morning, the guardian angels would watch and report. In after beer at night, into the Sip 'n' Sizzle, pool with the locals, laying challenge quarters up on the table edge, some of Blondena's mushrooms or onion chips, then out again with heat lightning on the midnight horizon, the guardian angels would watch and report. Lying in bed with all the windows open, heat thunder rumbling way off on the midnight horizon, you could hear them come up the gravel drive one, sometimes two, in the morning, singing at the bottom of the hill, reporting to one another what they'd seen in the middle of the darkness.

"The adult was sitting out on the highway!" one would say.

"They *always* sit out on the highway at night," another would say, then there would be some crunching of steps up the gravel for a while but no voices, sort of like they all would be thinking what it was like sitting out on the middle of highway 61 this time of year at one, sometimes two, in the morning. For the record, the adult didn't *always* sit out on the middle of the highway. Sometimes it was up on the reflector posts that marked the asphalt edge, sometimes it was up on the utility box just beyond the headlights, and sometimes it wasn't there at all. Off in the dark hunting, that would be the conclusion then. All the options, all the variations in nightly behavior, were dutifully passed around for all to discuss and remember. And so a hundred guardian angels went about their summer tasks until the little owls actually started coming out of the burrow. Then there would be electricity in the voices walking up that gravel in the dark. Then there would be electricity in little John's voice and movements coming out of town. And then also, was when it began to be obvious that if you're a burrow owl, then maybe even a hundred guardian angels are not enough.

The exact personality of a baby burrow owl is best left undescribed by me. The reason for this is because I know nothing about it. These are common birds but I've never handled a live young one, never hand-fed one, never felt talon or beak or gaze of a young burrow owl. Oh, I've done some other owls, seen the vicious terror of a captive great horned nestling, seen the almost comic bouncing-off-the-walls razor-bramble-balls-of-fluff screech owl babies, held the untold gentleness of a sun-dazed barn owl adolescent, straight out of that dark hole in Alligator Canyon not knowing what to do with three feet of wings, but I've never held a little burrow owl. Alive, that is. So their personality is left to the imagination. Nay, it is stronger than that. To be a guardian angel means you have an obligation to construct a personality for them. Right or wrong, your conclusions about baby owls' personalities are almost best imagined. So the owls' first gift to their guardian angels is a behavior that forces a human to use an imagination that may have lain idle for years, that may have atrophied amidst the sterile toxic learn-the-book vapours of much of today's higher education. Besides, there may not be enough human ingenuity in Nebraska to actually *catch* one alive. Secondly, the excitement of knowing you may be wrong, anticipating that one in a million chance to hold a baby burrow owl, is also an owl gift to guardian angels. But finally, their supreme gift to their guardian angels is an overpowering sense that parts of nature are so fragile that a human touch would destroy them. Later, for example, it was only in desperation, after a communal discussion of the best actions, that a decision was made for one person to actually walk up to the nest. Some gift. I hope you all get one some day. I hope some part of nature gives you that overpowering sense of *look but don't touch.* But back to baby owls' personalities.

They had a way of scrunching up their wings as they'd run back into the hole. They had a way of looking out through the sunflowers with the wariest of high curiosities. They had a way of scattering themselves over the burrow mound in a perfectly composed photograph, a photograph whose composition disintegrated into scurries when you tried to take it. They had a way of watching the traffic—Broncos, Wagoneers, hurtling junk with Colorado tags, Ford station wagons, all it ever took in anyone's mind to make it to the beaches of Lake McConaughy. They had a way of making you think they could see the *whole big* world just beyond their own flightlessness and could not wait to get out there in it, standing up on that hot burrow sand lip just flapping their little wings as hard as they could, determined flapping, hinting at the grace of an owl's wing, with imagined liftings of a little owl's body raised on toes with talons. So many times the guardian angels thought how that traffic on the curve north of town was about all of the whole big world those little owls saw, besides the sunsets, of course—they had a perfect view of the sunsets that summer. You could almost get the feeling those Sandhills sunsets were calling those little owls off into the whole big world of owl darkness, away from the burrow within a scurry's reach of safety. Then the guardian angels discovered what they didn't in a million years ever want to discover: little burrow owls also had a way of getting themselves killed in the traffic out on highway 61.

It was late morning and hot when they brought the first one in and laid it on the table—flattened, head smashed and oozing, ants everywhere, pitiful little wings back against the formica. But the feathers seemed still to have the essence of owl. Owl feathers are always soft, even the flight feathers, but especially the long down on the breast. You got the feeling right away there really wasn't very much burrow owl beneath all those feathers, that there really wasn't very much animal to make a decision that it was time to see the world and play in the traffic. Things got sad in a hurry, in a big way, but then things perked up a little bit. The first one was really too small to be out on highway 61. Maybe highway 61 was something that would not happen to the rest. Maybe the rest would fledge out over that cultivated row, would stick to the bar ditch until it melted out into real Sandhills with other badger holes away from the highway 61s, would maybe lift on those soft wings across all the highway 61s to the horizon their very first time. But all those "maybes" were human "maybes," the products of angels' hopes for things they could see but not touch. Then someone brought in the second of the burrow owls: flattened, oozing, ants everywhere, and not too long after that someone brought in the third, off highway 61, and suddenly those compositions seared so hotly in that imaginary photograph disintegrated into carrion out on highway 61. And all that day the guardian angels went about their work thinking that little burrow owls they had seen out trying wings, scurrying into that old badger hole, were now only carrion. And all those late nights up the gravel drive the conversation turned to the number left, the fading hope that one would make it, the possibility of simply blocking off highway 61, a college student idea to be sure, and brazen, but born of the atmosphere of Keith County. And it was way into August when I slammed on my brakes on that split lane curve, not giving a damn about traffic coming north on highway 61, and picked up the last of the smashed burrow owl young.

It had seemed like such a good idea when we, humans, had analyzed the nest site earlier in the summer. No one would stop to bother the nest on that particular curve. The choice was so obviously a case of that island of wilderness in the midst of civilization, with the civilization itself generating the isolation. It was not safe to stop a car on that curve north of town. Evidently, neither was it safe for an owl to stop and build a nest. I brought the last fledgling into camp and painted its picture. Painted pictures are for creatures that intersect my life in some emotional way, some way I want to remember. The owl was laid out on my torn leather work glove. It seemed like a good idea to paint it laid out on the glove instead of trying to reconstruct a live-looking owl from a specimen whose death rent the feelings of all those people. Then it seemed like a good idea to take the body out afterward and put it on the ant pile. Can't forget we're scientists, now, can we? If you put a specimen on the ant pile, the ants are supposed to turn that specimen into a perfect teaching skeleton. Some student from years into the future would learn something from that skeleton. I wondered how the label on that skeleton would read and what it would mean to that unknown future student.

The excitement of hope left the guardian angels then, but there was still an owl on the lip of that burrow. I am still after all these years of talking about birds not much of an ornithologist, for I personally could never decide whether it was a lonely adult waiting out the summer's end or a last older young, not so brave, perhaps, as the rest, with the uncertainties of highway 61. For those final days the exhilaration was simply gone. We watched silently that one owl sulking in the sunflowers and back in camp said little to one another. Then we all went home. It was over as quickly as that. There was no owl on the lip of that burrow on a hot Saturday afternoon as I turned the corner into town, Lido trailing. Little John strained to see, anyway. It was not particularly pleasant to

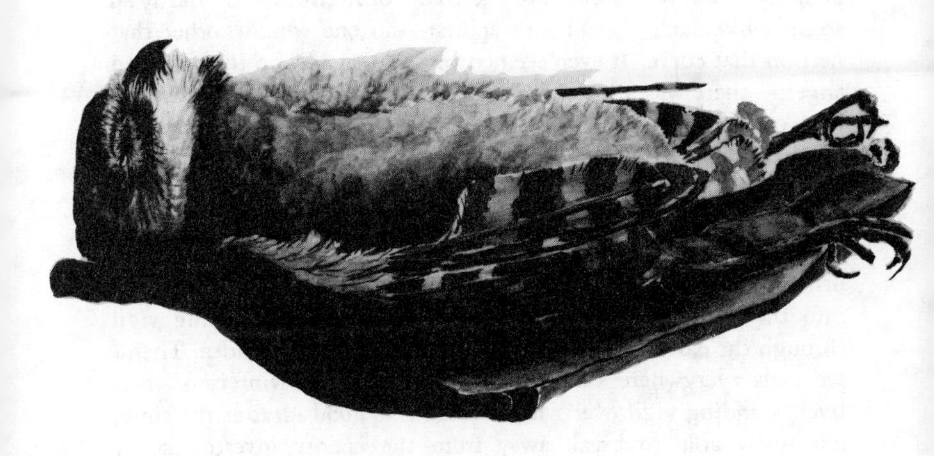

". . . it began to be obvious that if you're a burrow owl, then maybe even a hundred guardian angels are not enough."

leave Keith County that day. But then it's never been all *that* pleasant to leave.

The price of fledging was high, for those little owls, maybe too high to make it worth the first attempt at freedom. The price of fledging was high too for those parent owls, for that price included all they did to make that badger hole a home, the synthesis of eggs, the warmth given those eggs, the foraging for foods of increasing sizes those long days, even, thinking back, the energy expended on courtship, all those energies had to be included in the price of fledging a few young out into the traffic of highway 61. And it all seemed like such a good idea at first—no one would bother that nest on that curve. It even seemed like a great idea to the guardian angels—there was a lot of early talk about what a lesson in nature that owl nest would be for thousands of tourists that summer, tourists who maybe never even thought about an owl nest. But I never came away that summer with a good feeling about what seemed at first to be such a good idea. Instead, I had then, and still have, this vision in which sits that parent owl, up on the orange utility box out in some volunteer wheat, all through the fall and into the winter blowing drifts up around an owl standing vigil through the dark over what seemed at first to be a good idea. Then I see owls everywhere standing vigil through the winters of their lives, standing vigil over what seemed like good ideas at the time, not really able to break away from the energy investment, or migrate on to other good ideas.

Those visions were still in my mind when, a week later, and everyone gone back to the city, I once again went west into Keith County. It was the very end of August. The subject of my study was "disturbed areas." But there was an owl on the lip of the burrow, still standing on what seemed so long ago to be a good idea. In the end, no one had told that owl it was not a good idea to build a nest in *that* badger hole. The owls chose that hole freely. That wild freedom of choice had in turn given others so much: a summer of thrills, excitement, commitment, emotional investment in a single element of natural history, feelings and thoughts that would never have been felt or thought had that pair of burrowing owls not chosen to nest beside highway 61. But in the end, that wild freedom of choice had given others a stunning lesson in the fragility of wilderness.

Those kinds of lessons, especially when they stick with a person all his life, are not that easy to come by. It is impossible, nowadays, for me to remember the sadness of that owl nest without at the same time wondering if such a powerful lesson in fragility could ever have been taught to one human by another. Then I go on wondering and wondering what we as a community lose by making one another stay away from the badger holes of all highway 61s.

13

Disturbed Areas

TONY was catching grasshoppers. Once you've seen Tony catch grasshoppers, then you realize that grasshopper catching is best left to an expert.

"I went to the Thunderboat races one time, back when I was a kid," he said, then looked out over Arapaho Prairie. He had his own preferred net: small hoop, cloth bag except at the end where it was typical mesh. Tony said that kind lasted longer and I made a mental note to buy Tony's kind for students next summer. "Right here around the well is a disturbed area; much higher populations in these disturbed areas. Besides, you get the real pests in these places. Go over there in that clump of sage you'll find *Hypochlora alba,* almost restricted to the sage, with cryptic coloration. *Hypochlora alba* is one of my favorites; I may work on *Hypochlora alba*. Back over in these disturbed areas get a lot of *Melanoplus,* worst pests. Ranchers around here hate 'em."

"How were they?" When you get a couple of biologists out walking over Arapaho Prairie then a lot of times there is all this background of various insect talk, grass talk, hope, ambition, failures, and money talk, and you have to sort out the thread of conversation from all that.

"Oh, everyone ought to see them once. All that noise, the incredible speeds." He swept the net; grasshoppers roiled up over the prairie for a few yards. He stood for a time sorting through his catch, throwing most of it away, putting the right species for the moment in a plastic bag. Before the morning was over he'd go on to several more. Change his "search image," he'd say, so that net sweep would pick up a certain kind out of what looked to me like a million kinds all blowing in every direction. *"Mermiria bivittata*—slant face," he said. *Slant face* in grasshoppers is one of the most intriguing features of any animal. You'd think the zoos would be full of slant face grasshoppers.

"Where'd you see them?"

"Up in Seattle. We left early. I was too young to drink beer. *Spharagemon collarae."*

I asked him how to spell *"Spharagemon collarae." Spharagemon collarae* was to become one of my all-time favorites, sometimes with orange antennae, a little ridge up along their back, but most of all their mottled brown colors. The specimens were things of true beauty. But then most of the animals I've held have been things of true beauty, it's just that some are more so, like *S. collarae.* It's one of those that's so incredibly beautiful in its own way that you remember it. When you can go out into the field yourself, without Tony's help, and catch only *Spharagemon collarae,* then people say you've "learned a grasshopper."

"Yeah," said Tony, "I think you ought to come back out and see them once." He nodded in approval of what seemed like some frivolous self indulgence involving nearly a thousand miles of driving. Sometimes you need a neutral opinion to give approval of some idea like coming back to Keith County to watch unlimited hydroplanes, Thunderboats, make disturbed areas all over Big Mac. Sometimes you need a person who understands the biology of disturbed areas to give you approval for going into one, maybe even being a part of one, maybe even of creating one. There are some people in the world who are so totally sincere, so devoid of duplicity, so wonder-struck and open-minded, that their approval is all that is needed to validate *any* activity. Tony is one of those kinds of people. "Yeah," said Tony again, flicking his net over a disturbed area, "everyone ought to see them once." That was about all I needed.

The "Big Lake" has never been one of my major sources of intellectual stimulation. Oh, it's great for a party, a sun tan, but biologically it's pretty sterile. So I never spent much time up on the big lake. But you can't stay around these parts for even a few hours without hearing all about the big lake. You can't go to a social gathering without hearing how it was before they built the dam. Like most irrigation and flood control reservoirs, Big Mac's water level fluctuates too much to provide a really rich community of plants and animals. Shallow water over some land is potentially a fertile spot, but it gets disturbed and dried when the water goes down. No sooner does that disturbed area start to explode in prairie richness than the lake goes up and everything is flooded out. The lake itself, though, generated a gigantic disturbed area back in the 1940s. The disturbance gave rise to a whole new community of ideas, of opportunities, of lingering opinions, of a history fast fading into folk lore. Those disturbances from the 1940's had not died down when a committee decided to bring in the Thunder-boats.

The race course was marked off as a three mile rectangle out toward the west end of the big lake, over along the north side. It was a perfect viewing place: flat beaches below miles of sand cliffs. You could see them from a long way out there perched up on a cliff: big mobile home campers, boats attached, silhouetted against the haze over Ash Hollow, waiting for Thunder on Big Mac. The simple truth is that an unlimited hydroplane can turn a corner at a hundred miles an hour and never bank once, never depart from that flatness over the water, then scream out over the straightaway a bunch faster yet. That's one of the things I found out down where they had the race course marked off. The other thing I found out is that these boats don't leave any wake at top speed. Oh, all this rooster tail goes up, but then it all falls back down and there's no wake, at least at a hundred miles an hour. There's the noise, like Tony said, and the incredible speeds, but there is not much of a disturbed area in the *water* after that machine goes through. When they slow down, though, coming back into the pits, then there *is* quite a wake. I contemplated the meaning of that observation: when something happens fast enough there is no disturbed area. It's only when things slow down so you can sense them that they register. I adjourned twenty miles to the east for a beer just to find out if I could hear those airplane engines.

Twenty miles east of the race course is a place called Kingsley Cafe, that is really only part of Kingsley Lodge, which gives no one the impression of being a lodge. It's a gas station/grocery store/tackle shop/beer outlet. You can rent a boat or rent a motel room. The motel rooms are across the highway from Kingsley Cafe. There is a redwood deck along the back of the Cafe, but the building itself is old, frame, and hung on the edge of what was once a bluff of Brule limestone. There are yuccas all around the deck and a cottonwood tree comes right up through a hole built in the floor. At certain times the cotton comes down from that tree like snow. Then you can sit out on that deck about ten in the morning on one of those glassy days, watching the boats go out across Big Mac to some favorite spot, and talk about important things. It's such a wonderful place, none of my friends have ever been able to figure out why it seems to be a financial disaster.

We went up to Kingsley Cafe a few years ago, for example, just to have a beer and sit out and watch the lake. The Cafe had just changed hands. It was always changing hands back in those days. It will probably always change hands. Kingsley Cafe may be a permanent disturbed area. It changed hands more than usual that summer. The new girl showed up one day, probably just a couple of days before we went up there. She was full of stories about the lake, the first of a legion of stories. Probably not one of them was true, but you could tell a mile away she believed them all. You could almost tell she needed to believe all the stories for some reason of her own.

"I'm scared of that lake," she said. "Scared, really scared. I'd never go out there in a boat. You see people way out there in the middle in a little boat. You'd never get *me* out there in any boat. It's a hundred feet deep, more than that, maybe two hundred. There's abandoned cities down there, ghost towns, spooky!" She set the beers down on the table. The glasses were frosty under this particular management. "Railroad tracks, trains, houses, farms where people used to live, big trees, there's all that down under the lake. You know that big concrete thing? That's full of dynamite. Dam starts to break, they'll blow up all that dynamite and the water will go out onto McGinley's. They got dynamite planted all along the south shore; same reason. God is that scary! I lived here all my life. Went to one of those little schools up here in Arthur County. *Never* get me out on that lake!" She counted change. "My God, *water skis,* some people go out there on *water skis!*" Her eyes widened and her breathing was more regular, deeper, almost with a rattle in her throat. "Come to think of it, can't really figure out why I took a job so close to the lake. Maybe it's 'cause there's not all that many jobs for girls my age in Arthur County, 'less you live on some big ranch."

We laughed about the girl a few times after that. She no longer works at Kingsley Cafe. All that conversation took place before I came to believe the lake stories because they're just as good believed as true. We still go out there in boats. Bill Muncey was about to go out there in a boat called *Atlas Van Lines*. Maybe those lake stories make it all the better out there in a boat a hundred feet above towns and old trains. Lots of dead cottonwoods down under there, too. Those cottonwoods used to be right along the river and around farmsteads. People, the parents and grandparents of today's lake people, used to sit out under those cottonwoods and look out over grass. Bluestem, gramma, sand reed, switch grass, used to find all those down there before the Corps came along and disturbed it all.

Impoundments are never popular with everyone. Hundreds of families have to be bought out, condemned, moved to higher ground. In the Plains those people are the ones that lived next to the water: the choicest places. Their land was floodplain, often, recharged with silt, rich, alive in August, and their cattle were rarely thirsty. Such people never want to be moved. With the lake comes another kind of people: exploiters, people who never homesteaded, drifters, unlimited hydroplane drivers and crews, folks who never carried water in buckets for fruit trees a mile away back before they built the big dam. But impoundments are no different from other disturbances. The plow, the grazing steer pawing at sand around a well tank, the four-wheel-drive vehicles pawing up, down, and around the valley where unlimited hydroplanes will race, concession stands and rows and rows of rented portable toilets, prairie fire, they'll all bring in organisms who never homesteaded: drifters, exploiters. Sometimes these kinds are ones that spend their whole lives looking for disturbed areas. Sometimes they're the ones that make disturbed areas on purpose. These were my thoughts as I turned back out on that August blacktop west to where the Thunderboats lay under swarms of mechanics. Something's going on! The established order is disrupted! No plans for a drifter, just like to poke his nose into what's happenin', see if maybe there's an opportunity, now things is not so locked up! Opportunity for what? Who knows. I gravitated toward the biggest disturbance to be *planned* for Big Mac since they built the dam: The First and Last Annual Thunderboat Races.

Maybe before we get to the Thunderboats, I should throw in a short diversion here, a disruption in this tale of disturbances. There are people who are unable to keep their lives and thoughts organized. Their attention strays. Their minds are occupied with too many things. They are "overextended" say their bosses. But sometimes when a whole big bunch of things gets mixed up in their minds, they get a sense of some thread of meaning, some common driving principle, that unites all those ideas. They may even sit down at a typewriter and try to tell a story that illustrates an intangible act of synthesis skittering through their subconscious. Sameness bores these people; pathologically. Organization is the epitomy of sameness. The unordered disarray spawning their driving principle is orders of magnitude more exciting than organization. Their minds scream out, *"Don't you see? Can't you see??!"*

I am one of those people. My writings and thinkings are filled with disordered arrays. You are being dragged through this nutrient broth. Turn the page, go on to another chapter if you wish, throw down this book, check out something by John Updike. Don't feel compelled to ask what common thread of prairie life ties into meaning that random mixture of Tony, grasshoppers, Kingsley Dam, Kingsley Cafe, the girl from Arthur, Arapaho Prairie, all those different kinds of grasses, cocktail party talk on some redwood deck along the south shore, airplane engine mechanics, unlimited hydroplanes, the trampled clay of Spring Valley Road where *Atlas Van Lines* now creeps angled behind her tender through the quickening breeze of a Sandhills evening. Disordered arrays have a life of their own. They spontaneously generate syntheses. Such syntheses are coded for, perhaps, by relict human genes reaching across the eons making us social animals. Could it be the evolution of law, policy, regulation, organization, simplicity, is a systematic spewing of some Stone Age adaptation that sought the security of order? Could it be the human mind is allergic to random mixtures?

The evidence is not conclusive. Although we are no longer in the Stone Age, we are saddled, parasitized, even, by those nucleotide sequences that served so well dirty Neanderthal. Could it be this parasitic gene for security makes comforting unifying threads of meaning from randomness? *Disturbance represents opportunity.* Meaning! The ancient gene has struck! Disturbances leave tracks, unlike *Circus Circus* pink against the *terre verde* of Keith County at 104 mph, *disturbances leave wakes.* More meaning! The ancient gene has struck again! The traces of wakes can be followed through the years: the successions of Arapaho Prairie grasses, the bitterness of descendents who saw their fathers' homestead flooded where the randy *Squire Shop* crew, greasy, frets at whitecaps off the end of a pier. Did I begin this witches brew of a chapter knowing that ancient gene would strike from within to synthesize meaning from swirling seemingly disconnected subjects? Yes. Let's go back to Arapaho Prairie, do things slowly, get the principles straight, see what nature does with a disturbance, then return to the biggest planned disturbance in Keith County since they built Big Mac.

Arapaho Prairie is a research area, owned by The Nature Conservancy, managed by the University of Nebraska, and located southwest of Arthur. It covers two square miles. It was homesteaded in the early days, its virgin sod broken to satisfy the Federal government's requirement that homesteads be improved, and it's been grazed and burned since. But it's not been cultivated since the Great Depression. The soil is very sandy. Arapaho is two of twenty thousand square miles of such sandy soil and like the rest of that land, is thrown into dunes. It was hoped in the beginning that a study of Arapaho Prairie would reveal the nature of biological forces driving the Sandhills. Tony started studying grasshoppers, some others started classifying plants, marking home ranges of lizards, compiling bird lists. But the romantic of them all, Tyrone Harrison, came to Arapaho Prairie to study prairie grasses and disturbed areas. A poet's temperament, a prophet's vision, a martyr's commitment to the preservation of grasslands, an absolute disdain for academic housekeeping, those were the tools he brought to Nebraska for the study of Arapaho Prairie. I have known people who are so swept up in wonder that they distort the thinking of any who venture close. Their force fields are so powerful that they bend the gravity of state agency politics. Their electromagnetic waves block transmissions from any outside source. They are the teachers of all teachers. Most of the people I have known like this have been ornithologists. I have only known one person in my life who had these powers derived from a love of prairie grasses. His name is Tyrone Harrison.

Ty Harrison, like anyone who studies the prairies, is somewhat of a specialist in disturbed areas. His discovery that parts of

Arapaho had been cultivated in the early 1900s touched off an explosion of understanding of the distribution of grass clumps on those two square miles. He began identifying species, measuring roots, measuring rhizomes, measuring clumps, studying maps, talking with all the neighbors, taking verbal land use histories, taking photographs, drawing conclusions. His virulent amazement at his own discoveries swept with the heat of an outrageous epidemic through all who occupied the sterile cages of our city building. There was never a time in my twenty-plus years as a professional biologist that I ever once considered getting caught up in prairie grasses. But there was never a time in those twenty years that I became so consumed with another's ideas as on that day Tyrone Harrison began telling me of the succession of prairie grasses on the old plowed fields of Arapaho. Arapaho Prairie was abandoned as a farm forty years ago. Ty Harrison said you could use the observations from Arapaho to predict the rate at which Nebraska Sandhills prairie would reestablish itself following the failure of a farm. He claimed he could show even me, a person who had never taken a course in botany, the evidence right there on the prairie southwest of Arthur. The disturbed fields of Arapaho had been abandoned at almost the same time Kingsley Dam had been built. He claimed he could show me how it would take *eighty years* for the natural mixture of Sandhills grasses to reestablish itself after the disturbance.

"Meet me by the west well," I said. "Teach me the ways of prairie grasses."

Thus began my most basic lesson in the biology of disturbed areas, the basic lesson that guides my search for such areas within. Disturbances generate unholy cauldrons of hot synthetic brews, juxtaposed ideas. From such brews come things no one envisions. They are the sources of the truly new. I now generate them on purpose in my mind, mixing ideas, stirring vigorously, just to see if the stuff will explode. Sometimes it does. But back to the Pied Piper of all teachers: Tyrone Harrison.

*   *   *

There was never a person who walked the dunes in a more random manner. Yet there was never a person who made more sense in on-site conversations. I can't describe all this, really. We simply met at the west well and Ty Harrison got out of the truck and started walking toward the horizon. I ran after. He talked constantly. I wrote constantly; took pictures. He kneeled again and again to pull up roots, explain the anatomy, show how you could age a clump of grass by measuring the annual rhizome growth. Then he would stand and circle a whole clump with a gesture. Or he would stay on his knees and circle a whole clump with his cupped hands. His hands became a magic wand. One wave, one pass of a pointed finger, and things appeared out of a smear of sand green. Where before all had been the same, suddenly you could see clumps, half a dozen species, clones, the marching of bluestem across an old plow line, the furtive struggle of gramma, most loved of all the Gramineae. There were circles of all shades, all textures, scattered as might be blown seeds across the uniform haze of *Stipa*. The circles were each a single plant slowly expanding its territory in the matrix of needle grass. We struggled to the top of the highest bluff. He waved the magic wand again.

"Each circle is a clone, a single plant from a single seed. The largest ones came in very soon after the farm was abandoned." Of all the seeds that had been produced on that prairie since the 1930s, such a tiny fraction had survived to make a circle-clone! Uncountable seeds had been produced; I felt I could easily count the successes.

"It's been forty years since this place was abandoned," said Ty Harrison, "it will be another forty before the natural mix of Sandhills grasses is established on this old field." He was being generous. I had visions of us both being very old men, standing frail in the blowing heat of August forty years hence, and Ty Harrison saying in his old old voice: it's been eighty years since this place was abandoned, it will be another eighty before the natural mix of Sandhills grasses is established. And then I would think that when a disturbance creates opportunity, the wake lasts forever.

Somewhere in my red notebook there is a list of grass species, their local common names, notes of their places in the succession. For me to describe those species to you would be to waste the experience of being with Ty Harrison on Arapaho. I could never do the day justice. You would read, and thinking you'd been told how it was, would maybe be satisfied. But it can never be told how it was in the presence of such a teacher in such a place. Maybe the sounds of the common names will be enough: blowout grass, prairie sand reed, switch grass, sand bluestem, blue gramma. The list communicates "prairie" like no other list of words I know. Read that list aloud, alone in a dark room, and the reading takes you to the top of the highest bluff on Arapaho and the wind blows in your ears and the ghosts of homesteaders who made that plow line move slowly in the distance below. It may be the fact that on a natural Sandhills prairie you can find *all* those species, ideally in some kind of balance, that makes the list so impressive. A single name from the list doesn't take your imagination to the top of that bluff. But a single species from that list, blowout grass, is what you find, actually, at the top. The tops of Sandhills bluffs have a way of getting blown out down to the bare sand. You don't have to be told which species requires that most disturbed of Sandhills environments.

Deer rolling around, cattle pawing a trail, sometimes nothing more than the shape of a dune from another time channeling the wind, an ill-chosen pickup track, all can start the blowing. Next thing you know there's bare sand. The earless lizards move into the blowout edges then. Before long you get blowout grass, sparse, thin, blowing in circles making patterns in the sand. Sometimes you can stand in the bottom of one of those blowouts and look up into a whole concave hill of spaced out little circles where the wind's been whipping blowout grass. Somehow that blowout is a slate wiped clean. Somehow a slate wiped clean symbolizes opportunity; anyone can write their lives anew on a clean slate. There are no hierarchies to buck, to satisfy, to obligate you to things you didn't want to do today. Blowout grass is *genetically adapted to an assumed constant supply of clean slate.* It occupies the bare sand, stabilizes the blowout, and in a few years is replaced by less venturesome species. Then it moves on to the next disturbed area. Makes you sort of wonder what it would be like to be genetically programmed to a life that required a constant supply of clean slate.

Sociobiology is a relatively new field of science. Writings in this field claim certain human tendencies are genetically determined: mild polygyny, instinctive religion, fascination with weaponry, territoriality, dominance hierarchies, maybe some others. The *form* of these traits is culture, say the sociobiologists; the *fact* of them is genetic. Now I wonder, having been to Arapaho with Ty Harrison, if there are human equivalents of blowout grass, genetically programmed to a life that requires a constant supply of clean slate. Maybe Lonnie Paul Dinkle is one of those. Is he a distinct species? Are his genes so unique that he is of another kind altogether? And when such a mind-species occupies a disturbed area does it then start off a succession of mind-species: human needle grass, human sand reed, human bluestem? Does the succession always have to end with human blue gramma, the slowest to grow, lurking, drought resistant? Are Thunderboat drivers part of all this?

Karyn Stansbery is a stringer from the North Platte *Telegraph*. She is also a child of the Platte Valley, psychologically dominated by her awe of the land and sitting in the press tent. There has been a confrontation: Coors has supplied the beer and chicken for the press tent; Budweiser's public relations gal blisters a well-coiffured local committee woman for it—two women standing on a clay mud road bulldozed over and through the Spring Valley marsh exchange electrical charges, shots of acid, the wind is blowing, whitecaps scare giant thundermachines, womens' heels sink into gray mud, but Budweiser never gets its sign on the press tent. Local misunderstanding. Karyn Stansbery doesn't care. She's eating chicken supplied by Coors. The press tent is set up over prairie grass trampled by reportatorial feet. You can take off your hat inside the press tent, sit down in a lawn chair, talk about the wind and technical statistics of unlimited hydroplanes. Karyn Stansbery recites some technology. More people in western Nebraska have learned more about technology in the last month than they have learned in the last five years. Speeds, RPMs, weights, torques, spare parts, dollars, fuel mixtures, pressures on square inches of hull or fin—they've been told over and over again to little boys hanging around *Miss Madison*. She finally strikes a chord of meaning.

"Just think about it," says Karyn Stansbery, "this place used to be just some out of the way little park where nobody came. Now it will all be opened up when the thunderboats go. All these roads, the pits. Lots of people will be in here after the boats." Opportunity!

The initial opposition to the Thunderboats had proven somewhat ill-founded. The town itself was not swamped with uglies; no, in fact the town was too far away and there were not all that many uglies. Unlimited hydroplanes, it seems, draw a state fair type of crowd: salt-of-the-earth desirables. Furthermore, the Big Lake itself had proven too big to feel the biggest disturbance planned since the 1930s. One had to drive many miles out that undulating blacktop, way beyond Lemoyne, to get to the Thunderboat Races. It was only as one topped that last hill that the disturbed area became obvious. The truly big mobile home campers were in the choicest spots high on the bluffs along the race course. But the limits of the disturbance were not marked by campers. Instead, those limits were marked with sound. Dwarfing the most sumptuous vans were the speaker banks. There were two of them, a mile apart. They towered into the prairie wind, boxes upon boxes, held by wires. When they spoke, everyone within a mile listened. Most of the time they played country music and banter. The "race course was open" about thirty minutes total during the first two days. The prairie wind saw to that. Evidently a million dollars worth of hydroplane can't do what two thousand dollars worth of Lido can. Maybe that's being unfair. There was a lot of putting together of these models to be done. Not only that, *Miss Madison* was undergoing a sex change operation.

The committee that brought in the Thunderboats had been operating for a year. There were big bucks to be had, someone evidently thought. Since everybody in town was everybody else's friend, and so many people had great amounts of money sunk into the project, people opposed to the races had to find a way to wish their friends the very worst luck in the most benign manner. This was done by much sincere and public hope that the Thunderboat project would break even financially. On paper, then, the only thing lost would be committee time. So Ogallala spent a year or so trying to attract unlimited hydroplanes. This was a hunting area—geese, ducks, etc. The year-long flurry was not unlike that of hunters trying to attract geese and ducks out of the air. Set up the right conditions and maybe they'll come. The committee even set out decoys. An old retired *Miss Madison* angled on her trailer was hauled around the county, parked at truck stops, always with somebody in attendance to explain the technology to little boys. This *Miss Madison* would never race again, but evidently she worked well as a decoy. Before the end of August some other Thunderboats came, almost as if attracted after all, quiet, crews edgy, boats angled on their trailers, maybe suspicious, maybe not really knowing what to expect from a small town way out on the prairie. The new *Miss Madison* even came incognito—*Dr. Toyota* was painted on her side. Evidently town and Big Mac were not so threatening after all. *Circus Circus, Squire Shop, Kentuckiana Paving, Atlas Van Lines,* spent two days putting their engines in and fins on. *Miss Madison* underwent a sex change operation instead, neuter to feminine. *Dr. Toyota* was scrubbed off with paint remover. *Miss Madison* was painted back on. Then they were ready to race. The decoy had worked. It had attracted a small flock, one member of which was even the same species as the decoy!

There were briefings for the locals: dive teams, judges and other officials, rescue squads, paramedics, fire departments, announcers, ambulance drivers, all, it seems, with clipboards. Steve Reynolds ("one of the more popular drivers on the circuit") demonstrated the life jacket, helmet, other gear, for the paramedics ("if he's face down in the water I'd suggest you move a lot faster . . . don't *say* he's dead over the radio"). The spectator fleet crept closer ("all swimmers get out of the water so we can open the course"). Pit tour tram took one last sanitized pass through the sand and rocks of the pits, local pickup truck hauling, local mother doing the commentary by script ("the *Squire Shop* piloted by Chip Hannaur, one of the more promising young drivers"). But out of the way at the other end lay *Atlas Van Lines*—mean, outlaw lean, the bluest of Big Mac blue, a machine wedded to crusty Bill Muncey, all business, all fast, all design, a man's team among hopeful boys. Miss Thunderboat races, fair, freckled, an indoors girl, blistered and her high heels sank into the same loose sand that had earlier trapped the *Miss Budweiser* tender, the van with two or three airplane engines, the van with the machine shop inside.

A skyhook wandered over the field, finally selecting *Circus Circus* to try Big Mac, plucking tons of total speed up into the air while the officials' pontoon boat jostled the shore and the officials themselves, barefoot, pants rolled up to mid-calf, waded their twinkle toes into the biggest lake in cowboy country out to their barge. The crane set *Circus Circus* into the water. Steve Reynolds climbed in. Twelve cylinders of unmuffled airplane engine destroyed the Indian Summer and suddenly *Circus Circus* was way out there in the middle and all this water was up in the air behind her and she was going over a hundred miles an hour out against the sand green of bluffs way out around the end into the squinting reflecting sun, then screaming back at a hundred miles an hour, then around the corner at a hundred miles an hour, and then again and again came the thunder echoing off Big Mac and when Steve Reynolds finally cut that engine and *Circus Circus* made a wake at the end of her run, the crowd cheered and yelled at the fact that after the biggest planned disturbance on Big Mac had gone on for two days, finally one of the boats had made it out onto the water for ten minutes' worth of incredible speeds and incredible noise. And then was when it became obvious that the act of racing itself, the actual machine on the course, was only the excuse. The draw was not the boat on the water. The draw was the excitement of the whole business: the generation of a disturbed area.

In the end, *Circus Circus* didn't even qualify for the final event on Sunday. In the end, the Thunderboat Races broke even, except for the committee time, so plans for the next year were scrapped. The races were a good idea that worked like a charm—they just didn't make enough money. They didn't make enough money because committees are inherent parts of systems. Systems never produce what they're supposed to produce and always produce something else. In this case the system was designed to produce money. All types of accounting considered, all types of results evaluated, the system did not produce enough bottom line money. But it did produce something else: a disturbed area. It produced thirty thousand people who spent a lot of private time wondering how and why a little town like Madison, Indiana, should support an unlimited hydroplane on tour. It produced thirty thousand people who all of a sudden saw a lot of *kinds* of jobs being done for a living, jobs they never envisioned existed, jobs that kids in rural America may not always be told exist by a high-school counselor. It produced the grand opulence of American corporate wealth gone racing in the Nebraska Sandhills, *Miss Budweiser*'s tender stuck in the Spring Valley marsh, *Circus Circus*'s owner's mobile mansion parked where all could see and vice versa. Someone in that thirty thousand, maybe some kid otherwise destined to a bum's life, saw in that wealth the symbolic treasure of America to be had by the ingenious. That's a sight not often granted people in this part of the country. Maybe that sight struck a spark of ambition in some kid. So in the end the system produced an awareness of opportunity, which is not very surprising, considering it was a disturbance.

And just maybe, the personality of an unlimited hydroplane driver was the key element in this disturbance. After all, the Thunderboat Races were not really races, they were in fact three days of airplane engine mechanics and fifteen minutes of thunder. Was the excitement of being part of it more important than the race itself, and especially more important than the winner? Of course. And did the whole thing represent a grand demonstration of America's strength of opportunity, variety, wealth, ambition, technical sophistication, free competition? Of course. But would the whole thing have come off if someone had not been willing to climb into one of those monsters and drive the unlimited to its limits out across the waters of the biggest lake in all Keith County? Of course not. And would that same demonstration of opportunity occur in city after city if those drivers were not willing to do it again and again, making a living going fast, in place after place, knowing full well their very existence guaranteed a disturbed area? Of course not. And does this willingness to drive the unlimited to its limits produce thousands of people who can't remember the winner but *can* remember there's a place in this world for a mechanic, who *can* remember that in America an idea can be converted into reality, if only the reality of hull design and fuel mixtures, and who *may* reassess their own ambitions because of it? Of course.

It was time to go back three hundred miles east, past *Erma's Desire,* and think about what I'd been shown by disturbance. It had been a crazy week of mixed images—grass to hydroplanes, and all in between. Yet through it all there ran the thread of a different set of values. The people I'd dealt with—Tony, Ty Harrison, some pest species, boat drivers—were all people who saw more than discomfort in disturbance. These were all people who dealt constantly with opportunity, pure opportunity, the freedom to take advantage of it, and in some cases, as in some grasshoppers and some grasses, the requirement for a constant supply of it. Alone for miles in an automobile a person can do things with his mind, things that maybe other kinds of solitary confinement do not allow. After a week alone in Keith County, it seemed natural that a person could put together a tale of grasshoppers and Thunderboats, grass and Karyn Stansbery, Kingsley Cafe and *Atlas Van Lines,* and in the end come out with a common thread of meaning. When the synthesis finally comes, it is near Grand Island. I smile and turn the volume on my tape player *way* up. Disturbed areas represent opportunity. Too much order represents the opposite. When that kind of synthesis strikes about Grand Island, derived from original experience, I usually turn off at the rest stop and touch *Erma's Desire.*

14

The Experiment

"AMERICA was God's experiment." So said Saul Bellow, and I believe him. **ex.per.i.ment** *n.* 1. A test made to demonstrate a known truth, to examine the validity of a hypothesis, or to determine the efficacy of something previously untried. There is some reason this definition always comes to mind when I communicate with the world at large. I will explain why in the next few pages. That explanation comes from doing what I have done so often over the last many years: search for meaning, for truths, for principles of existence, among the seeingly simple observations of life among the wild. That search has taken the form of analogy, of a metaphor lived into convincing reality, of parable, of extensions of principles that govern the lives of lesser things, and finally of written expressions and watercolors stuck away in a closet or sold to the highest bidder. At the end of that search there lies the realization that I just may be a participant in the grandest experiment of them all.

Let's begin with a good talk about evolution. I won't defend evolution as a process; there is too much evidence from sophisticated journals of original science that demonstrates that populations of living things can change genetically with time. And that is the basic definition of evolution. But I know in addition that the word "purpose" has no place in discussions of evolution. Structures and functions are supposed to have arisen passively in response to "selective forces." We may not be able to justify use of *purpose,* but we can certainly ask what is the *proper function* of some structure. A lion's tooth, a whale's flukes, the eye of the vulture or the wing of the bat, all these things have "proper functions." Thus when these structures function in their proper way, they contribute most to the species survival *as a species.* These structures, their functions, and the species' ability to use them, are tested every day in the wilderness.

The pronghorn does easily what it does best—run, ghosting across the dunes, appearing and disappearing among the sand waves. The leg of the pronghorn was made to run. By whatever forces you credit with this synthetic feat, over whatever long years since the last glaciers inched down across what is now the Canadian border, that act of synthesis produced a hank of hair and stick of bone that was made to *run.* The turkey vulture, gracing crags of cowboy country, does easily what it does best—ride the wind. The vulture wing was made to sail. By whatever forces you credit with this synthetic feat, over whatever long years since Kansas was no longer an ocean, that act of synthesis produced an ugly bundle of feathers and hollow bones that could catch the wind like no other structure known or made by man. *Catch the wind!* And so it goes with the wild things. They all use their inherited attributes to the best of their ability to survive, as individuals, and as species. Those same forces of nature that made running machines, that built the best of all sails, made another device with another function. You know what that device is—the human brain. Just as the pronghorn leg runs best, just as the vulture wing sails best, this last structure of all the brains made for all other species, *thinks* best. It synthesizes; it creates; it spins its own thoughts, thoughts none other can take away. On with the experiment.

A test made to demonstrate a known truth. Is it indeed a known truth that of all things it does, the human brain best synthesizes, creates, spins thoughts none other can take away? Yes. And is Keith County, U.S.A., humanity's grand test to demonstrate that known truth? I think so. And can I extend those thoughts to ask, is it just possible that the *concept* of the United States of America is also perhaps humanity's grand test to demonstrate that known truth? Yes, I think so. Read on.

What has mankind invested in this local condition of relative intellectual freedom? A disproportionate share of its fossil fuel, a disproportionate share of mineral and metal resources. And what has been the return on mankind's investment? Futurists read the patterns of employment in the United States, compare those patterns to similar ones in other developed nations, then categorically state that development leads away from industrialization and toward a society whose raw material is *information*. Over half the so-called workers in this country work in information processing, use, retrieval, or generation jobs. They are not workers, they are thinkers. Even our play becomes play of the mind. Examples are too numerous to cite: professional football coaches don't work, they think; theirs is an information job; they create it, analyze it, act upon it, teach the others to do the same. Bookkeepers, bankers, secretaries, bureaucrats, insurance salesmen and -women, real estate agents, the list of neighbors' livelihoods marches down the road to infinity, but few touch raw ore, few wield a hammer at the forge. America, the world's schoolroom, has metamorphosed. Reviled by Iranians, disdained by Europeans for lack of restraint, but possessing in relative abundance that rarest and most critical of resources—free thought, America's metamorphosis may only be the expected result of the grand test to demonstrate a known truth. The known truth is that humans differ from other species in the complexity of the brain. The Experiment demonstrates that given the same opportunity as the turkey vulture, the pronghorn, *our* species will assume an existence that demands it do best with its unique structure, that places a premium on creative thought, intellectual skill, survivorship of the mind.

In any scientific laboratory, an analogous result would direct further research, dictate future methodologies. A cell biologist would optimize the culture conditions for a cell line. No cell biologist would strive for her cells' survival, *in their best functional condition,* and purposefully leave some vitamin out of the culture fluid. It would only be for manipulative purposes that cells would be kept without some essential nutrient. Perhaps a restriction on the free uptake of some essential nutrient would prevent the cells' metamorphosis. Perhaps some poison chemical would inhibit the cells' ability to *contribute* to the culture broth, thus preventing metamorphisis within the culture. Perhaps if those cells were allowed to metamorphose into their best functional condition, the scientist would lose his or her ability to manipulate them! Maybe there is a touch of scientist, of cell biologist, in certain humans around the world. Maybe there is a touch of would-be scientist in political aspirants here in the Grand Experiment. Maybe our metamorphosis is one of the reasons it is so difficult to be President. Maybe the *opportunity* to be intellectually free, to compete on the open market with thoughts, to act in concordance with freely spawned ideas, individually, is that essential nutrient necessary for metamorphosis. Maybe our metamorphosis is only the expected metamorphosis in the life cycle of that species we call "human."

Now there's an interesting idea. Maybe without intellectual freedom, populations of *Homo sapiens* are doomed to remain in arrested development, their ontogeny hung up in midrecapitulation of their phylogeny, their actions more reminiscent of animal ancestors than of extant congeners. Maybe the maintenance of a vegetative existence without concurrent intellectual opportunity *guarantees* arrested development. Maybe the cost of education, on top of the cost of food and shelter, is only the investment that guarantees that the investment in food and shelter will not be wasted in an animal existence. Maybe the cost of such education is only the price we pay to become *human,* to enter the metamorphosed state. Maybe when you think of it that way, education is not so expensive.

* * *

A test made . . . to examine the validity of a hypothesis. Master Experimenters design their work around *null hypotheses.* Null hypotheses assume there is no difference between things. Statistically, *no* difference is easier to disprove than difference. Thus even when answers are known or strongly suspected, Master Experimenters will phrase a null hypothesis as a working device. But herein lies the secret to "successful" science: choose *the* hypothesis out of all possible to test. Then at every turn, with each disproof, appear two more approaches, so that even an unwise first choice might eventually lead, through an interconnected set of choices, to the key null hypothesis that proves all. But where to begin? *Null Hypothesis:* there is no difference between *Homo sapiens* and the other species of this planet. *Null Hypothesis:* there is no difference between Keith County (even a mental symbolic one) and the other places of the planet. *Null Hypothesis:* there is no difference between *Homo sapiens* in Keith County (even a mental symbolic one) and *Homo sapiens* in the other places of this planet (even the mental symbolic ones).

The differences between humans and the lower animals have been documented; no need to continue here. The first Null Hypothesis is disproven. The mental metamorphosis described above provides the disproof. America has been the test tube of history. Keith County has been only one of many detectors, a geographic spectrophotometer, through which the chemical reactions in that test tube have been "seen."

The differences between Keith County, even a mental and symbolic one, and the other places of this planet, have been well documented; not much need to continue here. The second Null Hypothesis is disproven. We began this journey, you and I, with a pass along Interstate 80 near Grand Island and a spell of contemplation of abstract sculpture. That journey took us into the times and places unhindered. We saw the freedom of animals, a freedom we imposed on their wild lives with our interpretations. We thought the free thoughts that *Erma's Desire,* with her points, spurred us to think. We talked with the Corfields and the Packards.

We seined the chub, killed the sucker together. We contemplated the wild call of the five string banjo and the northern oriole. Vehicles, fish, and rivers untamed, rebellious companions, we tethered ourselves to the prairie wind and we watched the baby birds live and die. You've met my friend Lonnie Paul Dinkle. We've even endorsed disturbance. Thank you for your company. Your trip to the place I teach biology has taken you away from your present life and obligations; they have not been able to reach you out north of Ogallala. Your experiences have been those lived by hundreds of the University of Nebraska's Cedar Point Biological Station students, by faculty members paid from taxes in the Sovereign State of Nebraska. You've gone back to college. Go back, now, to whatever you were doing. Take some Keith County with you. Dip into it when you need some, when you need to think unfettered thoughts and feel nondirected feelings, when you are ready to free that original and creative element of your own self. The second Null Hypothesis will forever be disproven.

The difference between humans in a symbolic or mental Keith County and humans in other mental or symbolic places, have perhaps not been well described. Perhaps those differences have not been identified. Perhaps what we are really asking is what has mankind produced as a result of the *chance* to think a free thought? What has mankind produced as a result of the *opportunity* to chase a dream, to pursue curiosity to its end? Some possible answers follow:

The notebooks of Leonardo da Vinci, the mathematics of Einstein, the stories of John Cheever, the Winslow Homer watercolors, the George Beadle (of Wahoo, Nebraska) Nobel Prize-winning experiments in fungal genetics, the observations of Al Wallace, and the writings of Charles Darwin, the gardens of Linnaeus and those of his father. That about spans it. The X-ray crystallography of James Watson, the cut marble of M. Buonarroti, those are more of the same. There are historians who would say all these great works have been the products of single free-thinking minds. There are historians who would say that only the works of individuals survive the millenia and appreciate. As an unrelenting biologist, I would go further: In this species, *Homo sapiens,* blessed with the brain of brains, no single mind has produced an animal contribution that has stood the test of time. Great military commanders have swept the face of our planet, but do their conquest lines remain intact and valuable as the words of Homer? Hitler's new boundaries lasted less than ten years. What is it that survives? It is the art, the music, the literature, none produced by committee, none painted or carved by politician. American school children play music by Bach, Beethoven, then graduate into higher grades to study Goethe. No American school child studies the letters of Joseph Paul Goebbels.

But I can't draw, you say; I never really learned to play a musical instrument, you lament; my creative powers are nothing compared to the great ones of all history, you conclude, with little supporting evidence. Besides, I don't have time for all that, I must work to make a *living*. With that phrase you seal the conversation shut. It is clear you have not found your own Keith County within, that cowboy country place to retreat into when the trophic pressures of society exercise only your animal talents. You've never met your own Lonnie Paul Dinkle, or if you did, his actions and attitudes disturbed you beyond words and you were unable to see his desperate message conveyed by the paralanguage of his approach to life. You've never waded into your own Mudhole #1 or seen the grebes upon it, if you are unable to see the extremes among the commonplace, the vulnerability of love displayed along the road, the safety of that vulnerability when it is displayed to romantics who refuse to allow society to exercise only their animal talents.

My father and mother used to say, "Get an education, son. That's something no one can take away from you, your education." So I went to college and got an education. It led me at the grand age of forty to a cabin in the hills along the south shore of Keystone Lake, a place the maps don't even recognize. The education I received at the University of Oklahoma never began to match the education I received in that cabin in the hills. But no one at OU ever told me I could not *attempt* to enter graduate school in zoology with only a bachelor's degree in mathematics! I was told in essence, "Here are the courses, pay your tuition. Come to class and try if you want to. If you succeed, great; if you fail, tough " I didn't know it at the time, but I had just been introduced to the Keith County within. It didn't seem strange then, it only seemed normal, that no barriers should be placed by a society, an organization, between me and my desires of curiosity.

But today it doesn't seem strange, it only seems normal, that those barriers exist everywhere, between curiosity and the opportunity to explore it. Lonnie Paul Dinkle would never be admitted to the Graduate College of the University of Nebraska, yet he is out there doing pioneer research on the nature of human soul. He's found somewhere his Keith County within. Erle Corfield would never come to the Graduate College of the University of Nebraska, even if admitted. Yet there are thousands of Ph.D.s who will never experience the rich range of human existence the way Erle will, and already has. Although he lives in Arthur County, he's found his Keith County within. Mark Safarik may continue his profession of part-time bartender, hocking gold buttons for next week's groceries, but it is not his degree from the University of Nebraska that lightens your burden, brightens your day. It is the metaphors he spins when he retreats into his own Keith County.

And none of the thirty or fifty scientific papers I have written, the committee reports I have generated, or the thousands upon thousands of dollars of your tax money I have spent on research will ever touch your life the way my books will. Written on whimsy, jam-packed with ideas no formal scientific publication would ever look at twice, spun out page by rebellious page in the early morning hours of my basement office, they are the product of a retreat into that Keith County within. Total individuality, the action of a single mind doing something not considered normal or even a part of the job society thinks I am trained for, behavior that causes the powers untold worry, those are the things one finds in that inner Keith County. The alarm rings. No, it doesn't ring, it cuts the darkness as rough electronic static. I move the lever, turn it off, but am on my feet. It is three hours after midnight. Coffee is dripped; the cup sits beside the Selectric. Recycled paper is reeled into place, and out comes *Keith County Journal, Yellowlegs,* this tome, whatever its name, hundreds of pages of history and philosophy that will never see the light of day. It is not difficult to disprove the last Null Hypothesis: there is no difference between a human in a mental or symbolic Keith County and a human in another place. The last Null Hypothesis is of all the easiest to disprove.

Our experiment so far is "successful." We have demonstrated known truths, examined the validity of hypotheses, but have yet to *determine the efficacy of something previously untried.* What is previously untried is a wholesale and widespread retreat into Keith County within.

Someone is going to read this and say I am proposing anarchy. No, anarchy has been tried, there is no need to test it. The efficacy of anarchy was tested in Africa several million years ago and anarchy was shown not to work. As a result, the human genome now contains powerful instinctual drives for organization, leadership, weaponry, hierarchy, religion, those civilized versions of the traits that allowed a naked, clawless, four-foot ape man to compete with lions and his own relatives out on the plains of Serengeti. Anarchy gets *Australopithecus* starved into extinction. The genes for social organization, however, set his foot upon the road to man. But what you may not find in all texts is recognition of the wilderness within which ape-man became organized, religious, stratified into dominance hierarchies, equipped with the latest bone club. *Australopithecus* had no need to seek a Keith County within, he lived in a perpetual state of Keith County. Now millions of years later our one identifiable act of civilization has been to discard the wilderness in which our regulatory drives *are themselves regulated.*

Unregulated regulatory drives kill the human brain, the human spirit, and sap the creativity that separates us from the lion.

I can't help thinking so often of Lonnie Paul Dinkle's question up there in the dark on top of that bluff we call Cedar Point. "Don't you just feel you *need* what's out there, Doc?" "What's out there" is miles of grass, Mudhole #1, and all like it, western grebes, tiger beetles, killifish, Arapaho Prairie, disturbed areas, a place to sail my Lido. I *perceive* what's out there as wilderness. Certainly it is *relative* wilderness. Certainly when I'm out there in it, my regulatory drives are themselves strongly regulated! Yes, Lonnie Paul, I feel I *need* what's out there. Whatever that is is probably what drove me to become a biologist in the first place. That's why I've put so much effort into preservation of the wilderness within.

That's also why I believe with unquestioned faith those individuals who say the *sight* of wild fields is essential for a healthy mind. These people of course are simply searching for a straw, any straw, to help barricade against the onslaught of "economic" development, the mining of mountain tops, the irrigation canals one can see from Mars, transbasin diversion of Platte water into the Little Blue. All other reasons for preserving the marsh prove weak against the Public Power and Irrigation District. The labyrinthine cabalae of agency politicians defy all reason as I stand for the last time, glazed eyes sinking inside at the places where there used to be wrens. And yet there are some who, failing to stop the hydroelectric plant, simply say: *humans must have the sight of wilderness in order to keep a healthy mind, to remain a human.* What irony, that in final desperation, reason failing, they turn to the strongest but weakest reason of all. What irony that the strongest reason of all is the least defensible in terms of dollars. What irony, that our source of humanness is in the hands of those who require terms in dollars! What convoluted emotions govern us, that a few acres of cattails could come to symbolize the world of opportunity that greeted *Australopithecus!* And what unquestioned faith leads me to say *I am human;* I must have that freedom and wildness symbolized by the marsh. If I can't have that wildness at my door, I will have it within. My regulatory drives will never go unregulated!

I am ending this set of essays now. This book contains my concluding remarks about Keith County, Nebraska. By the time this volume hits the shelves, it will be the eighth year of our work in the Sandhills. Eight years is enough. In those years there has been time to think, time to paint, a time to experience the rebirth of a man who, at the height of his career, finds he has strayed from the wilderness that brought him to that peak. Keith County has symbolized values forgotten in the crush of civilization. It has provided the simple wonder that drives a person to study nature in the first place. And, it has shown us how much a part of the wild world we actually are—physically, mentally, but most of all, emotionally. In the final analysis, that wilderness is more than Arapaho Prairie, more than the physical presence of the South Platte, more than a fragile dozen acres that can be swept away forever by eminent domain, a price "paid" for the priceless. In the end that wilderness is the freedom to think your own thoughts, to put them into practice, to have them judged with the same impartiality that greets the fledgling swallow.

It is dark now as I pull off the interstate into the Grand Island rest area. It is one of those lingering warm evenings in spring upon the northern plains. Semis are idling; car doors are slamming; there is talk from the dark, and the brittle crunch of boots on volcanic ornamental gravel. Off down the sidewalk a cigarette brightens then returns to a dull mark. A dog chain rattles; a voice reassures the animal, then goes on with some conversation. I am drawn out into the darkness beyond the buildings; the dew is gathering and my boots are wet. The little lake is suddenly there, reflecting traffic a mile away. *Erma's Desire* is taller than I am, her angled points move out into the dark above. The lights catch them, then move away; someone is going. Even at this distance she is protean, even in the dark she has the power to compel the unique interpretation, the thought no one else could generate, the thought no one else could take away. Alone in the dark we touch; she is old, rusting, cold; the sound of my knuckles rings hollow up through a spire. It is the spire that points back toward the organization. We touch again; she is warm but rough, the sound of my knuckles comes back a symphony of echos up through the spire that blocks my view of blackness above. It is the spire that points west to Keith County.